T0271026

Wastewater Treatment in Steel Industries

Treatment of toxic wastewater generated from different unit operations of the steel industry is a matter of extensive research. The most important is the utilization of different treatment techniques, especially ozonation, electrocoagulation, precipitation, and various hybrid methods for the remediation of toxic pollutants. This book covers case studies of various treatment techniques utilized for the remediation of different steel industry unit operation wastewater. The book is aimed at researchers and graduate students in wastewater treatment and chemical engineering.

Features:

- Focuses on case studies of steel industry-generated wastewater treatment.
- Discusses different unit operations of the steel industry along with wastewater generation.
- Provides insights for the remediation of toxic industrial wastewater by different advanced treatment techniques.
- Considers the environmental impacts of the generated steel plant wastewater.
- Includes focused knowledge and future insights on wastewater treatment.

Wastewater Treatment in Steel Industries
Case Studies, Advances, and Prospects

Mihir Kumar Purkait, Piyal Mondal,
Pranjal Pratim Das, and Deepti

CRC Press
Taylor & Francis Group
Boca Raton London New York

CRC Press is an imprint of the
Taylor & Francis Group, an **informa** business

Designed cover image: © Shutterstock

First edition published 2024
by CRC Press
2385 NW Executive Center Drive, Suite 320, Boca Raton FL 33431

and by CRC Press
4 Park Square, Milton Park, Abingdon, Oxon, OX14 4RN

CRC Press is an imprint of Taylor & Francis Group, LLC

© 2024 Mihir Kumar Purkait, Piyal Mondal, Pranjal Pratim Das, and Deepti

ISBN: 978-1-032-41615-1 (hbk)
ISBN: 978-1-032-43223-6 (pbk)
ISBN: 978-1-003-36626-3 (ebk)

DOI: 10.1201/9781003366263

Typeset in Times LT Std
by KnowledgeWorks Global Ltd.

Contents

About the Authors

Dr. Mihir Kumar Purkait is a professor in the Department of Chemical Engineering and head of the Centre for the Environment at Indian Institute of Technology Guwahati (IITG). Prior to joining as faculty in IITG (2004), he received his Ph.D. and M.Tech in chemical engineering from Indian Institute of Technology, Kharagpur (IITKGP) after completing his B.Tech and B.Sc. (Hons.) in chemistry from University of Calcutta. He has received several awards. He is the director of two incubated companies. He is also technical advisor of Gammon India Ltd and Indian Oil Corporation, Bethkuchi for their treatment plant. His current research activities are focused on four distinct areas, viz. (i) advanced separation technologies, (ii) waste to energy, (iii) smart materials for various applications, and (iv) process intensification. In each of the areas, his goal is to synthesize stimuli responsive materials and to develop a more fundamental understanding of the factors governing the performance of the chemical and biochemical processes. He has more than 20 years of experience in academics and research and published more than 200 papers in different reputed journals. He has 8 patents and completed 24 sponsored and consultancy projects from various funding agencies. Prof. Purkait has guided 18 numbers of Ph.D. students.

Dr. Piyal Mondal received his B.Tech in chemical engineering from the National Institute of Technology Durgapur, West Bengal (India) in 2012. He completed his Master's and Ph.D. degrees in chemical engineering from the Indian Institute of Technology Guwahati (India). His research is dedicated to preparing various surface-engineered nanomaterials and polymers for specific environmental applications. His research is also focused on the synthesis of polymeric membranes, green synthesized nanomaterials, and hybrid techniques to combat wastewater treatment. He has fabricated different prototypes for environmental separation applications. Currently, he has authored six reference books. Moreover, his publication comprises 17 peer-reviewed articles in reputed international journals, with several more under review. He has presented more than 15 papers and received several awards in poster and paper presentations in his field at international and national conferences.

Pranjal Pratim Das received his B.Tech (2014) and M.Tech (2017) degrees in Food Engineering and Technology from Tezpur (Central) University, Assam, India. He is currently pursuing his Ph.D. degree in Chemical Engineering from the Indian Institute of Technology Guwahati, Assam, India. His research work is purely dedicated to industrial wastewater treatment via electrochemical and advanced oxidation treatment techniques. His research focus is also related to the application of integrated electrochemical and oxidation processes for the treatment of specific unit operations of steel industry effluents. He has extensively worked in the treatment of cyanide and phenol contaminated wastewater of Tata Steel Industry, Jamshedpur, India. He has published several peer-reviewed research paper in reputed international journals, along with patents and book chapters. He has fabricated and demonstrated many pilot plants for the green energy generations from different wastewaters. He has worked widely on various iron and steel making industry effluents and has also delivered many pilot-scale set-ups to several water treatment facilities across the state of Assam, India for the treatment of toxic contaminated wastewaters.

Dr. Deepti pursued her Ph.D. in the Centre for the Environment from the Indian Institute of Technology, Guwahati (IITG). She received her Master's degree (2016) in industrial pollution control from Manipal Institute of Technology, Manipal, Karnataka, and B.Tech (2014) from Alvas Institute of Engineering and Technology, Moodbidri, Karnataka. Her current research activities are focused on advanced separation technologies, conversion of industrial waste to valuable resources, wastewater treatment, and industrial effluent treatment technologies. Moreover, she has a patent, and her publication consists of four peer-reviewed articles in reputed international journals, with several more under review. She has presented more than four papers and received awards in paper presentations in her field at international and national conferences. She has been closely working with steel plant wastewater from Tata Steel and has published several peer-reviewed articles on this.

1 Introduction to Steel Industry

1.1 INTRODUCTION

The industrial sector's need for water has grown significantly over the past few decades, leading to increased exploitation of water resources to meet demand. In addition, the concentration of various pollutants in groundwater and surface waters has increased due to the unregulated discharge of industrial wastes (Purkait, Sinha, Mondal, & Singh, 2018a, b, c). To reduce the negative impacts on human health and the environment, effective treatment of industrial effluents is required before discharge (Mondal & Purkait, 2017; Sontakke, Das, Mondal, & Purkait, 2021). The iron and steel industries are essential to the growth and development of economies in developing nations like China and India. Iron and carbon are the main components of steel, which is one reason why they are frequently referred to as a single substance and why iron is necessary for the production of steel (Rawat, Srivastava, Bhatnagar, & Gupta, 2023; Samanta et al., 2021). Iron ore is the primary metallic component utilized in "primary" steel manufacturing, whereas scrap steel is the primary metallic component used in "secondary" steel production. Nevertheless, these distinctions can be difficult to make because "waste is frequently employed in primary fabrication, and iron is frequently utilized in electric furnaces, which are the usual unit for secondary steel fabrication". India is a rapidly populating nation in Asia, with China's population predicted to pass India's by the middle of the 2020s and a more than sixfold increase in GDP since the early 1990s (Mallett & Pal, 2022). With 5% of worldwide production, India is the second biggest fabricator of steel in the world. According to estimates, India will produce 227 Mt, 347 Mt, and 489 Mt of crude steel by 2030, 2040, and 2050, respectively, up from its 2019–2020 production of roughly 111 Mt. In India, the quantity of steel manufactured via different methods can vary from 77% of the volume for Integrated Steel Plants (ISPs) to 60% for small-scale steel manufacturing plants utilizing induction furnaces. Globally, steel fabrication through various methods averages around 80% of its capacity. Following attempts to liberalize the economy in the 1990s, steel manufacturing in India truly took off (Kim et al., 2022). Since 2010, it has doubled, and as shown above, it is predicted to rise further during the following decades. After China and the United States, India is the third biggest purchaser of completed steel goods.

Blast furnaces (BFs) are the key component of the steelmaking production process in India. They produce an intermediate carbon-saturated form of iron. The hot metal is then typically fed into a basic oxygen furnace (BOF) to produce liquid steel, followed by cooling and casting into ingots. It is then heated, chilled

DOI: 10.1201/9781003366263-1

into sheets, and finally transformed into steel products. Direct reduced iron (DRI), which is the process of reducing the oxygen content of iron ore by mixing it with fuel and a reductant, is another approach to creating primary steel (Mahieux, Aubert, & Escadeillas, 2009). Sponge iron is sometimes used to describe the solid form of iron that results from the DR process of reducing iron ore to iron. India is also the greatest producer with an estimated 34.15 Mt of DRI production. The most common fuel used worldwide is natural gas, whereas in India, 81% of DRI is generated in coal-based rotary kilns and the remaining 6% is formed using gas-mediated methods, viz. coke oven gas, natural gas, gas from Corex process and syngas from coal gasification. The DRI is then put in an electric furnace, either an induction furnace (IF) or an electric arc furnace (EAF), along with some scrap steel (IF). Less heat is needed when manufacturing steel in an EAF furnace with scrap as the primary source, which is a benefit (hence more energy efficient) (Das, Prakash, Reddy, & Misra, 2007). However, as was already said, steel manufac-tured in India utilizing the DRI procedure is more carbon intensive because less scrap is used compared to iron ore and coal. Additionally, it's critical to remem-ber that steels created via the IF approach have lower load-bearing qualities and cannot be used in several situations. Nearly 50% of India's total steel fabrication is produced in electric furnaces. Approximately 28 Mt and 27 Mt of steel were produced by EAFs and Ifs separately. The steel industry in India has a variety of production methods. Steel production is divided into three categories, according to the percentage of each type of production process: IF (27%), EAF (28%), and BF-BOF (45%) (Manso, Polanco, Losañez, & González, 2006). Figure 1.1 depicts a schematic representation of iron and steel manufacturing processes.

The coke oven is the integrated route's most water-intensive component, using 0.5 m^3 of water for every ton of steel fabricated and causing the largest wastewa-ter release. Additionally, each ton of ingot steel a BF produces requires 7.6 m^3 of water. The common pollutants found in coking wastewater include suspended particles, COD, ammoniacal nitrogen, oil and grease, and other contaminants like phenol, CN^-, and SCN^-. Water is primarily used for furnace cooling in steel-manufacturing (EAF and BOF) processes, which leads to a significant quantity of suspended solids (1000–5000 mg/L) with improved thermal load. Before being released, the coke oven wastewater is typically treated by biological oxidation in steel plants (Das et al., 2021a; Purkait, Mondal, & Chang, 2019; Purkait, Singh, Mondal, & Haldar, 2020). Toxic contaminants cannot be success-fully removed with this method, though. Wastewater from finishing processes like hot and cold rolling comprises scales and debris loads of 100–200 mg/L and oil concentrations of 10–25 mg/L (Das et al., 2021b). Other finishing processes like acid pickling and electroplating release metallic ions into the effluent and leave behind residual acids, respectively. Coking wastewater contains a sizable number of very hazardous pollutants, such as benzene (1500–6000 mg/L), PAHs (30 mg/L), and heterocyclic aromatic compounds (400 mg/L) (Bu, Li, Cao, & Cao, 2020).

This chapter primarily focuses on the process description and unit operations in the steel industry, viz. blast oxygen furnace, electric arc furnace, ladle furnace

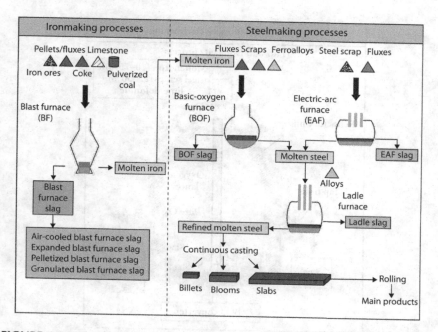

FIGURE 1.1 Flowchart of iron and steel manufacturing processes. (Reproduced from Yildirim & Prezzi, 2011 @ Elsevier.)

refining, continuous casting, and rolling mills. Also, detailed descriptions of the slags generated from the iron manufacturing and steel-furnace processes have been discussed and summarized.

1.2 PROCESS DESCRIPTION AND UNIT OPERATIONS IN STEEL INDUSTRY

1.2.1 BASIC OXYGEN FURNACE

The molten iron manufactured in the blast furnace and steel wastes are fed into basic oxygen furnaces (BOFs), which are housed at combined steel mills alongside a blast furnace (BF) (Das et al., 2021c). The ideal BOF charge typically consists of 80–90% molten iron and 10–20% steel scrap. The BOF charge contains steel scraps, which is crucial for cooling the furnace and keeping the temperature at 1600°C to 1650°C so that the necessary chemical reactions can occur (Rahmatmand et al., 2023). Figure 1.2 shows a simplified illustration of a BOF. Steel scrap is added to the furnace first, and then, immediately after this charge, a crane is used to dump a ladle of molten iron (about 200 tonnes) above the steel scrap. The charge is then blasted with 99% pure oxygen at supersonic speeds by an oxygen lance dropped into the furnace. Intense oxidation reactions eliminate the charge's contaminants throughout the blowing cycle, which lasts for 20–25 min. The temperature increases between 1600°C and 1700°C when carbon dissolved

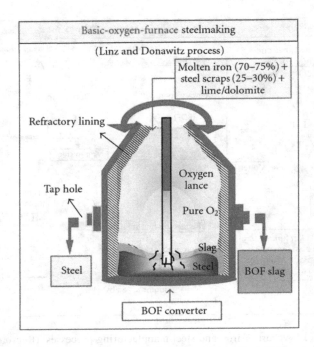

FIGURE 1.2 Schematic diagram of basic oxygen furnace (BOF). (Reproduced from Yildirim & Prezzi, 2011 @ Elsevier.)

in the steel burns to produce carbon monoxide. As a result, the scrap is melted, and the iron's carbon content is decreased (Juckes, 2003). During the cycles of oxygen blowing, the furnace is also charged with fluxing components, viz. dolomite or lime in order to eliminate the undesirable chemical components of the melt. Impurities react with burnt lime or dolomite to generate slag, which lowers the concentration of unwanted components in the melt. Toward the end of the blowing cycle, samples of the molten metal are taken, and their chemical makeup is examined. The oxygen lance is removed from the furnace once the desired chemical composition has been reached. The slag from the steel manufacturing procedure floats on top of the molten steel. To tap the steel into ladles, the BOF is slanted in a single direction (Poh, Ghataora, & Ghazireh, 2006; Samanta, Das, Mondal, Bora, & Purkait, 2022). The steel made in the BOF can either be transferred to a continual caster in which semi-finished shapes such as slabs, billets, and blooms are solidified in combined steel plants, or it can go through additional refinement in a secondary refining plant. The BOF is slanted once more in the reverse direction to dump the liquid slag into ladles once all the steel has been taken from it. The slag produced during the steelmaking process is later treated, and the resultant material is known as BOF slag (Changmai et al., 2020; Samanta, Das, Mondal, Changmai, & Purkait, 2022). The chemical makeup of the BOF slag is established by the chemical processes involved in impurity elimination (Shen, Wu, & Du, 2009).

1.2.2 ELECTRIC ARC FURNACE

Instead of using gaseous fuels, electric arc furnaces (EAFs or mini-mills) generate the heat required to melt recovered steel scrap and turn it into standard steel. Since the primary feedstock for the EAF steel manufacturing procedure is steel scrap with some pig iron, it is not reliant on the output from a blast furnace. Graphite electrodes are used in EAFs, which resemble large kettles with an odd notch or spout on one side. The EAF's roofs include pivot and swinging capabilities to make it easier to load raw materials. In scrap yards, steel scraps are separated, graded, and classified into various steel classes, whether they are heavy melt or in shredded shape. To ensure that the melting circumstances and the chemistry of the final steel in the furnace are within the desired range, different types of scrap are carefully put into scrap baskets according to their size and density (Forsido, McCrindle, Maree, & Mpenyana-Monyatsi, 2019). Electrodes made of graphite are then lowered into the furnace. When an arc is hit, electricity flows via the metal's electrodes and into the arc. The heat is produced by the electric arc and the metal's resistance to the electrical current. The electrodes are pushed deeper into the layers of debris as it melts. In some steel factories, oxygen is additionally injected during this procedure via a lance to decrease the scrap size (Yildirim & Prezzi, 2011). As the melting procedure advances, a pool of liquid steel is produced at the furnace's base. Either the scrap is added to the furnace along with CaO in the form of dolomite or burnt lime, or CaO is blown into the furnace during melting. The refining metallurgical operations, viz. dephosphorization, and decarburization, are conducted after several baskets of scrap have melted. Through an oxygen lance, oxygen is pumped into the molten steel throughout the steel refining procedure (Forsido, McCrindle, & Monyatsi, 2020). During the oxygen injections, some iron is oxidized together with other impurities in the heated metal, such as aluminum, silicon, manganese, phosphorus, and carbon. Slag is created when these oxidized substances react with lime (CaO). Carbon powder is inserted into the slag phase floating on top of the molten steel during the refining process, which causes carbon monoxide to develop. The slag foams as a result of the carbon monoxide gas created, which improves the efficiency of the thermal energy transfer. The EAF is slanted, and the slag and steel are tapped out of the furnace into separate ladles once the desired chemical composition of the steel has been attained. A ladle is used to move steel to a secondary steel fabricating section for additional refinement. Ladles or slag pot carriers are used to transport the molten slag to a slag processing facility. Up to 300 ton of steel can be produced in an electric arc furnace every cycle (a cycle takes 1–3 hours to complete). Initially, only high-quality steels were produced using the EAF steelmaking method because it was costlier than the BOF procedure (Nair, Mathew, A R, & Akbar, 2022).

1.2.3 LADLE FURNACE REFINING

Steel fabricated by the BOF or EAF procedures can be further refined to achieve the necessary chemical constituents once the initial steelmaking operations are

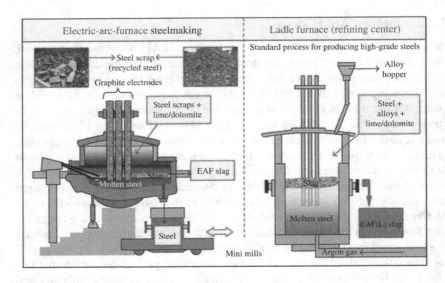

FIGURE 1.3 Schematic diagram of electric-arc-furnace and ladle refining processes (Reproduced from Yildirim & Prezzi, 2011 @ Elsevier).

finished. Secondary steelmaking operations are what these refining procedures are known as. Refining procedures frequently produce high-grade steels. Final desulfurization, degassing of oxygen, nitrogen, and hydrogen, degradation of contaminants, and final decarburization are the primary purposes of secondary refining operations. Molten steel created by the EAF and BOF processes may go through the aforementioned refining steps, depending on the desired level of steel quality. For secondary metallurgical processes, the majority of micromills and combined steel plants have ladle-furnace refining sections (Zhou, Wu, Wang, Wang, & Dong, 2013). Figure 1.3 depicts a simplified diagram of an EAF and a ladle furnace refining machine.

Three graphite electrodes are also included in ladle furnaces, which resemble scaled-down versions of EAF, and heat the steel using an arc transformer. Argon gas is typically pumped through a conduit at the bottom of a ladle furnace to stir and homogenize the liquid steel inside. The sulfur content of steel can be reduced to 0.0002% by inserting desulfurizing chemicals via a lance. During deoxidation, silicon and aluminum are added to create the oxides alumina and silica, which are later absorbed by the slag produced during the refining procedure (Wang et al., 2022). Additionally, the appropriate alloys are introduced to the molten steel to accurately alter the steel's chemical composition through an alloy hopper that is attached to the ladle furnace in order to make varying standards of steel. Prior to the start of casting operations, ladle furnaces serve as a storage space for the steel. Ladle furnaces offer greater operational flexibility and lower the cost of producing high-grade steel (Zhou, Zhu, & Wei, 2021).

1.2.4 CONTINUOUS CASTING

A short, water-cooled vertical copper mold is used to continually feed liquid steel into it while simultaneously withdrawing the frozen shell with the liquid steel it holds. Continuous casting technologies are utilized to solidify around 50% of the world's supply of liquid steel. Complementing the liquid steel flow into the mold with the speed of the strand's withdrawal from the mold is the primary control parameter for continuous casting. A tiny, refractory-lined supplier known as tundish is positioned above the mold to draw the steel from the furnace ladle and control the flow rates. Driven rolls, which contact the strand once it has formed a thick, hardened shell, control the withdrawal rate. Similar to the apparatus used in ladles, a stopper rod or sliding gate regulates the feeding of the caster mold from the tundish. The grade of the steel determines the temperature "window" within which the liquid steel should be kept in the tundish. By pouring steel via refractory tubes, which are submerged in the metal, or via broad sleeves, which are pressured with argon, shielding can be achieved. An argon cap is frequently placed on top of the lidded tundish (Craig, Camisani-Calzolari, & Pistorius, 2001).

Copper was used to make the mold due to its excellent heat conductivity. In order to prevent the solidified shell from adhering to its walls, it oscillates up and down while being substantially water-cooled. Additionally, oil or slag kept on the steel meniscus and flows into the space between the mold and strand is used to lubricate the mold wall. When in use, casting powder is continuously added to create the slag layer. In addition to lubricating, it prevents air from coming in contact with the liquid steel, serves as a heat barrier, and sops up impurities. Many continual casters have sensors in the mold that naturally match the pace of the strand removal with the flow of liquid steel into the mold. The strand is quickly water-cooled by spray nozzles as soon as it leaves the mold, with a shell thickness of only about 10 mm (Cui & Luo, 2017). The maximum casting speed at this point is determined by the strength and soundness of the shell since breaching it would create a breakout of liquid steel and cause harm to the caster. Several rolls support the strand as it descends to prevent the shell from bulging under the liquid steel's ferrostatic pressure. The supporting rollers get bigger and farther apart as the shell thickness rises at the end of the cooling zone. The secondary cooling zone is also frequently referred to as the metallurgical length. It can range in length from 10 to 40 m, depending on the cross-section of the strand and the casting pace. Computers are frequently used to control and automatically alter the water flow to the numerous nozzles in the various sections as casting circumstances vary (Wang, Deng, Dong, Shi, & Zhang, 2012). After passing through the final pair of support rolls, the strand enters the run-out table and is cut by one or more oxyacetylene torches while still traveling through the table. Commercial continuous casters are constructed with several design philosophies. A straight mold, a straight secondary cooling zone, and vertical strand cutting are arranged vertically for some steels and solidification patterns. Other casters similarly use a straight mold and a vertical secondary chilling zone, but they bend the strand in a horizontal direction as it descends and cut it on a run-out table once it has set.

Everything is on one radius or numerous matching radii, all the way down to the straightener. In order to cast strands with various cross-sections, various design approaches are used (Sui, Zhang, & Zhang, 2022). A beam blank caster creates enormous, dog-bone-shaped pieces, that are directly send into a H-beam or an I-beam rolling mill, bloom casters solidify sections of 300 by 400 mm, and billet casters solidify 75–175 mm squares. With production rates of up to three million tonnes annually, massive slab casters stabilize sections that are up to 230 mm thick and 2550 mm broad. Some billet casters have six molds next to one another in a line, and they're all fed from the same tundish (Klimeš et al., 2020).

1.2.5 ROLLING MILLS

Steel is produced using the typical metalworking technique of rolling. Multiple rollers are used to pass a sheet of steel through them. While some rolling procedures only require two rollers, some require four or more. Regardless, the steel acquires a uniform thickness and consistency as it moves through the rollers. However, rolling mills are frequently used by steelmaking sectors to do the rolling. A rolling mill is a device that, as its name implies, is made to roll and manipulate sheet metal. At least one set of rollers can be found in rolling mills. They are utilized by steelmaking enterprises to alter the physical characteristics of raw sheet metal, including steel sheet metal (As'ad & Demirli, 2010; Das, Sharma, & Purkait, 2022). Sheet metal will undergo a rolling mill process, changing its physical characteristics. Multiple rollers are used in rolling mills to alter the physical characteristics of sheet metal. Sheet metal is compressed and squeezed when it passes the rollers in rolling mills. Consequently, the sheet metal may have had an irregular size or shape prior. But as the sheet metal goes through the rolling mill's rollers, it takes on a regular size and shape. The fact that rolling mills can support both cold rolling and hot rolling processes is significant. Cold rolling describes the process of pressing metal sheets between rollers at or close to room temperature (Lu, Sun, Wei, Li, & Zhang, 2021). On the other hand, hot rolling is the process of pressing a metal sheet between rollers at a temperature higher than the metal's recrystallization point. Hot rolling is a quicker and easier rolling method that enables more efficient production operations, whereas cold rolling enables the development of stronger and more lasting steel products (Bhattacharya, Mishra, Poddar, & Misra, 2016). Both cold rolling and hot rolling can be done in rolling mills. Furthermore, the following categories of rolling mill rolls can be separated according to the requirements of the process and arrangement of the rollers (Özgür, Uygun, & Hütt, 2021):

Two High Rolling Mills

It comprises two rollers that rotate anticlockwise to move the workpiece in the desired direction. The work item is fed between the rollers, which exert a strong force on it and tend to deform it into the required shape. Looking for mill rolls manufacturers can help you determine which two high-rolling mills is the most reliable and suitable for your needs. The two high-rolling mills are further separated into two types: high non-reversible machines, in which the work item can only be supplied in one direction, and the rollers revolve in one direction. On the other hand, the second high-reversible machine has rollers that revolve in both directions (Thompson & Si, 2014).

Three High Rolling Mills

Three parallel rolls are stacked one on top of the other in a roll stand that makes up the Three High Rolling Mills. The material is passed between the top and middle rolls in a single direction and the bottom and central rolls in the opposite direction by rotating the neighboring rolls in the opposite direction. The workpiece is rolled on both the forward and the backward passes. The workpiece travels between the middle and top rollers, then through the bottom and intermediate rolls. Roll mills of the highest caliber are offered by numerous steel roll manufacturers to satisfy every imaginable industrial need (Wang, Yang, & Sun, 2015).

Four High Rolling Mills

A roll stand with four parallel rolls stacked, one on top of the other, can be found at the Four High Rolling Mills. The top and bottom rolls function in opposition to one another. The top and bottom rolls, usually referred to as backup rolls, are larger than the two in the middle.

Tandem Rolling Mills

A set of two or three roll strands is arranged in parallel alignment in a tandem rolling mill. With the shift in the material's orientation, a continuous trip through each one might be feasible. Numerous mill roll producers offer high-quality tandem rolling mills to a variety of sectors (Cao et al., 2018).

Cluster Rolling Mills

A cluster rolling mill is a rolling mill with a first four-high structure in which each working roll is supported by two or more larger rolls for rolling heavy materials. Sometimes it could be necessary to use work rolls with a small diameter (Takahashi, 2001).

1.3 CONCLUSION

A significant quantity of sludge and slags are being produced from different unit operations, viz. basic oxygen furnace, electric arc furnace, and ladle refining processes of the integrated steel plants. The effective processing of crude iron ore in the blast furnace to produce molten iron is a very crucial step in the steelmaking process, which is then transported to a basic oxygen furnace and electric arc furnace to form molten steel, followed by ladle refining to produce high-grade steels. Slags are classified depending on the furnaces from which they are produced. The major types of slags that are formed from the iron and steel manufacturing plants are categorized as follows: (i) blast furnace slag (ironmaking slag) and (ii) steel furnace slag. The steel furnace slag can be further categorized as basic oxygen furnace slag, electric arc furnace slag, and ladle refining slag. In addition, the continuous casting process can successfully produce semi-finished shapes such as slabs, billets, and blooms which are solidified in combined steel plants. At the end of the process, rolling mills (hot and cold rolling) offer uniform thickness and consistency to the steel sheet metal and impart the desired mechanical property to the finished products.

REFERENCES

As'ad, R., & Demirli, K. (2010). Production scheduling in steel rolling mills with demand substitution: Rolling horizon implementation and approximations. *International Journal of Production Economics*, *126*(2), 361–369. https://doi.org/10.1016/j.ijpe.2010.04.027

Bhattacharya, D., Mishra, A., Poddar, G. P., & Misra, S. (2016). Case study of severe strip breakage in rolling mill of thin slab casting and rolling (TSCR) shop of TATA steel, Jamshedpur. *Case Studies in Engineering Failure Analysis, 5–6*, 15–22. https://doi.org/10.1016/j.csefa.2015.11.002

Bu, Q., Li, Q., Cao, Y., & Cao, H. (2020). A new method for identifying persistent, bioaccumulative, and toxic organic pollutants in coking wastewater. *Process Safety and Environmental Protection, 144*, 158–165. https://doi.org/10.1016/j.psep.2020.07.022

Cao, J., Chai, X., Li, Y., Kong, N., Jia, S., & Zeng, W. (2018). Integrated design of roll contours for strip edge drop and crown control in tandem cold rolling mills. *Journal of Materials Processing Technology, 252*(September 2017), 432–439. https://doi.org/10.1016/j.jmatprotec.2017.09.038

Changmai, M., Das, P. P., Mondal, P., Pasawan, M., Sinha, A., Biswas, P., & Purkait, M. K. (2020). Hybrid electrocoagulation–microfiltration technique for treatment of nanofiltration rejected steel industry effluent. *International Journal of Environmental Analytical Chemistry, 102*(1), 1–22. https://doi.org/10.1080/03067319.2020.1715381

Craig, I. K., Camisani-Calzolari, F. R., & Pistorius, P. C. (2001). A contemplative stance on the automation of continuous casting in steel processing. *Control Engineering Practice, 9*(9), 1013–1020. https://doi.org/10.1016/S0967-0661(01)00089-2

Cui, H., & Luo, X. (2017). An improved Lagrangian relaxation approach to scheduling steelmaking-continuous casting process. *Computers and Chemical Engineering, 106*, 133–146. https://doi.org/10.1016/j.compchemeng.2017.05.026

Das, B., Prakash, S., Reddy, P. S. R., & Misra, V. N. (2007). An overview of utilization of slag and sludge from steel industries. *Resources, Conservation and Recycling, 50*(1), 40–57. https://doi.org/10.1016/j.resconrec.2006.05.008

Das, P. P., Anweshan, A., Mondal, P., Sinha, A., Biswas, P., Sarkar, S., & Purkait, M. K. (2021a). Integrated ozonation assisted electrocoagulation process for the removal of cyanide from steel industry wastewater. *Chemosphere, 263*, 128370. https://doi.org/10.1016/j.chemosphere.2020.128370

Das, P. P., Anweshan, & Purkait, M. K. (2021b). Treatment of cold rolling mill (CRM) effluent of steel industry. *Separation and Purification Technology, 274*, 119083. https://doi.org/10.1016/j.seppur.2021.119083

Das, P. P., Mondal, P., Anweshan, Sinha, A., Biswas, P., Sarkar, S., & Purkait, M. K. (2021c). Treatment of steel plant generated biological oxidation treated (BOT) wastewater by hybrid process. *Separation and Purification Technology, 258*(P1), 118013. https://doi.org/10.1016/j.seppur.2020.118013

Das, P. P., Sharma, M., & Purkait, M. K. (2022). Recent progress on electrocoagulation process for wastewater treatment: A review. *Separation and Purification Technology, 292*(April), 121058. https://doi.org/10.1016/j.seppur.2022.121058

Forsido, T., McCrindle, M., & Monyatsi, M. I. (2020). Application of EAFDS/ Lime integrated system for the removal of Cu and Mn from industrial effluent. *International Journal of Water and Wastewater Treatment, 6*(3), 1–6. https://doi. org/10.16966/2381-5299.164

Forsido, T., McCrindle, R., Maree, J., & Mpenyana-Monyatsi, L. (2019). Neutralisation of acid effluent from steel manufacturing industry and removal of metals using an integrated electric arc furnace dust slag/lime process. *SN Applied Sciences, 1*(12), 1–6. https://doi.org/10.1007/s42452-019-1649-z

Juckes, L. M. (2003). Transactions of the Institutions of mining and metallurgy : Section C the volume stability of modern steelmaking slags the volume stability of modern steelmaking slags. *Mineral Processing and Extractive Metallurgy, 112*(3), 177–197. https://doi.org/10.1179/03719550322500370

Kim, J., Sovacool, B. K., Bazilian, M., Griffiths, S., Lee, J., Yang, M., & Lee, J. (2022). Decarbonizing the iron and steel industry: A systematic review of sociotechnical systems, technological innovations, and policy options. *Energy Research and Social Science, 89*(February), 102565. https://doi.org/10.1016/j.erss.2022.102565

Klimeš, L., Březina, M., Mauder, T., Charvát, P., Klemeš, J. J., & Štětina, J. (2020). Dry cooling as a way toward minimisation of water consumption in the steel industry: A case study for continuous steel casting. *Journal of Cleaner Production, 275.* https:// doi.org/10.1016/j.jclepro.2020.123109

Lu, X., Sun, J., Wei, Z., Li, G., & Zhang, D. (2021). Effect of minimum friction coefficient on vibration stability in cold rolling mill. *Tribology International, 159*(December 2020), 106958. https://doi.org/10.1016/j.triboint.2021.106958

Mahieux, P. Y., Aubert, J. E., & Escadeillas, G. (2009). Utilization of weathered basic oxygen furnace slag in the production of hydraulic road binders. *Construction and Building Materials, 23*(2), 742–747. https://doi.org/10.1016/j.conbuildmat.2008.02.015

Mallett, A., & Pal, P. (2022). Green transformation in the iron and steel industry in India: Rethinking patterns of innovation. *Energy Strategy Reviews, 44*(February), 100968. https://doi.org/10.1016/j.esr.2022.100968

Manso, J. M., Polanco, J. A., Losañez, M., & González, J. J. (2006). Durability of concrete made with EAF slag as aggregate. *Cement and Concrete Composites, 28*(6), 528–534. https://doi.org/10.1016/j.cemconcomp.2006.02.008

Mondal, M., Mukherjee, R., Sinha, A., Sarkar, S., & De, S. (2019). Removal of cyanide from steel plant effluent using coke breeze, a waste product of steel industry. *Journal of Water Process Engineering, 28*(October 2018), 135–143. https://doi.org/10.1016/j. jwpe.2019.01.013

Mondal, P., Samanta, N. S., Meghnani, V., & Purkait, M. K. (2019). Selective glucose permeability in presence of various salts through tunable pore size of pH responsive PVDF-co-HFP membrane. *Separation and Purification Technology, 221,* 249–260.

Mondal, P., & Purkait, M. K. (2017). Green synthesized iron nanoparticle-embedded pH-responsive PVDF-co-HFP membranes: Optimization study for NPs preparation and nitrobenzene reduction. *Separation Science and Technology, 52*(14), 2338–2355.

Nair, A. T., Mathew, A., A R, A., & Akbar, M. A. (2022). Use of hazardous electric arc furnace dust in the construction industry: A cleaner production approach. *Journal of Cleaner Production, 377*(March), 134282. https://doi.org/10.1016/j. jclepro.2022.134282

Özgür, A., Uygun, Y., & Hütt, M. T. (2021). A review of planning and scheduling methods for hot rolling mills in steel production. *Computers and Industrial Engineering, 151*(July 2020), 106606. https://doi.org/10.1016/j.cie.2020.106606

Poh, H. Y., Ghataora, G. S., & Ghazireh, N. (2006). Soil stabilization using basic oxygen steel slag fines. *Journal of Materials in Civil Engineering, 18*(2), 229–240. https:// doi.org/10.1061/(asce)0899-1561(2006)18:2(229)

Purkait, M. K., Sinha, M. K., Mondal, P., & Singh, R. (2018a). Interface Science and Technology, Elsevier, Pages 115–144.

Purkait, M. K., Sinha, M. K., Mondal, P., & Singh, R. (2018b). Ch 2 - pH-Responsive Membranes, Singh, Interface Science and Technology, Elsevier, Volume 25, Pages 39–66, ISBN 9780128139615.

Purkait, M. K., Sinha, M. K., Mondal, P., & Singh, R. (2018c). Chapter 4 - Photoresponsive Membranes, Editor(s): Mihir Kumar Purkait, Manish Kumar Sinha, Piyal Mondal, Randeep Singh, Interface Science and Technology, Elsevier, Volume 25, Pages 115–144, ISBN 9780128139615.

Purkait, M. K., Mondal, P., & Chang, C.-T. (2019). Treatment of Industrial Effluents: Case Studies, CRC Press, 1st Edition, ISBN 9781138393417.

Purkait, M. K., Singh, R., Mondal, P., & Haldar, D. (2020). Thermal Induced Membrane Separation Processes, Elsevier, ISBN 9780128188019.

Rahmatmand, B., Tahmasebi, A., Lomas, H., Honeyands, T., Koshy, P., Hockings, K., & Jayasekara, A. (2023). A technical review on coke rate and quality in low-carbon blast furnace ironmaking. *Fuel*, *336*(December 2022), 127077. https://doi.org/10.1016/j.fuel.2022.127077

Rawat, A., Srivastava, A., Bhatnagar, A., & Gupta, A. K. (2023). Technological advancements for the treatment of steel industry wastewater: Effluent management and sustainable treatment strategies. *Journal of Cleaner Production*, *383*(August 2022), 135382. https://doi.org/10.1016/j.jclepro.2022.135382

Samanta, N. S., Banerjee, S., Mondal, P., Anweshan, Bora, U., & Purkait, M. K. (2021). Preparation and characterization of zeolite from waste Linz-Donawitz (LD) process slag of steel industry for removal of Fe^{3+} from drinking water. *Advanced Powder Technology*, *32*(9), 3372–3387.

Samanta, N. S., Das, P. P., Mondal, P., Bora, U., & Purkait, M. K. (2022). Physico-chemical and adsorption study of hydrothermally treated zeolite A and FAU-type zeolite X prepared from LD (Linz–Donawitz) slag of the steel industry. *International Journal of Environmental Analytical Chemistry*, *104*(3), 1–23. https://doi.org/10.1080/03067319.2022.2079082

Samanta, N. S., Das, P. P., Mondal, P., Changmai, M., & Purkait, M. K. (2022). Critical review on the synthesis and advancement of industrial and biomass waste-based zeolites and their applications in gas adsorption and biomedical studies. *Journal of the Indian Chemical Society*, *99*(11), 100761. https://doi.org/10.1016/j.jics.2022.100761

Shen, D. H., Wu, C. M., & Du, J. C. (2009). Laboratory investigation of basic oxygen furnace slag for substitution of aggregate in porous asphalt mixture. *Construction and Building Materials*, *23*(1), 453–461. https://doi.org/10.1016/j.conbuildmat.2007.11.001

Sontakke, A. D., Das, P. P., Mondal, P., & Purkait, M. K. (2021). Thin-film composite nanofiltration hollow fiber membranes toward textile industry effluent treatment and environmental remediation applications: Review. *Emergent Materials*. https://doi.org/10.1007/s42247-021-00261-y

Sui, Y., Zhang, H., & Zhang, X. (2022). Study on the creep behavior and microstructure evolution of a low alloy steel in continuous casting processing. *Materials Science and Engineering A*, *838*(February), 142828. https://doi.org/10.1016/j.msea.2022.142828

Takahashi, R. (2001). State of the art in hot rolling process control. *Control Engineering Practice*, *9*(9), 987–993. https://doi.org/10.1016/S0967-0661(01)00087-9

Thompson, S., & Si, M. (2014). Strategic analysis of energy efficiency projects: Case study of a steel mill in Manitoba. *Renewable and Sustainable Energy Reviews*, *40*, 814–819. https://doi.org/10.1016/j.rser.2014.07.140

Wang, L., Deng, C., Dong, M., Shi, L., & Zhang, J. (2012). Development of continuous casting technology of electrical steel and new products. *Journal of Iron and Steel Research International, 19*(2), 1–6. https://doi.org/10.1016/S1006-706X(12)60051-X

Wang, X., Yang, Q., & Sun, Y. (2015). Rectangular section control technology for silicon steel rolling. *Journal of Iron and Steel Research International, 22*(3), 185–191. https://doi.org/10.1016/S1006-706X(15)60028-0

Wang, X., Ni, W., Wei, X., Zhang, S., Li, J., & Hu, W. (2022). Promotion effects of gypsum on carbonation of aluminates in medium Al ladle furnace refining slag. *Construction and Building Materials, 336*(April), 127567. https://doi.org/10.1016/j.conbuildmat.2022.127567

Yildirim, I. Z., & Prezzi, M. (2011). Chemical, mineralogical, and morphological properties of steel slag. *Advances in Civil Engineering, 5*, 116–125. https://doi.org/10.1155/2011/463638

Zhou, Y., Wu, L., Wang, J., Wang, H., & Dong, Y. (2013). Alumina extraction from high-alumina ladle furnace refining slag. *Hydrometallurgy, 140*, 14–19. https://doi.org/10.1016/j.hydromet.2013.08.007

Zhou, Y., Zhu, R., & Wei, G. (2021). Application of submerged gas-powder injection technology to steelmaking and ladle refining processes. *Powder Technology, 389*, 21–31. https://doi.org/10.1016/j.powtec.2021.05.003

2 Classification, Sources, and Composition of Steel Industry Wastewater

2.1 INTRODUCTION

Water is extensively used for a multitude of activities, including descaling and dust cleansing, in addition to cooling operations in steel manufacturing. The procedures used to make steel use water from a variety of sources. While seawater is typically utilized for cooling, followed by pretreatment, fresh water is generally used for various other processes, including direct and indirect cooling. Three different processes are used to make steel: the integrated cycle, which uses virgin raw materials; the electric route, which uses scrap to melt steel in an electric arc furnace; and the open hearth process (Changmai et al., 2022; Khunte, 2018).

An integrated steelwork typically uses 28.6 m^3 of fresh water and typically releases 25.3 m^3 of wastewater for every ton of steel produced, although the average intake and discharge for the electric route are 28.1 m^3 and 26.5 m^3 per ton of steel, respectively (Deepti, Sinha, Biswa, Sarkar, Bora & Purkait, 2020b). Since evaporation accounts for the majority of water loss, the overall water consumption is modest (1.6–3.3 m^3).

Due to evaporation, the salt concentration in the seawater circulating system increases, for which seawater must be cooled and desalinized as it is harmful to the environment and also to plant equipment. Phosphates, for example, possibly promote eutrophication of the aquatic environment, whereas chlorides and sulfates can corrode equipment, and carbonates cause scale formation in pipes, leading to increased energy usage (Colla et al., 2016).

Many researchers observed the substantial formation of suspended solids during the processes, which is relevant to their water utilization features. Some of the research considered iron and steel industry effluents were highly toxic and unutilized components, necessitating mitigation (Purkait, Sinha, Mondal, & Singh, 2018a, b, c). The levels of chemical oxygen demand, suspended solids, electrical conductivity, and pH in converter dust removal sewage were also investigated (Purkait, Mondal, & Chang, 2019; Purkait, Singh, Mondal, & Haldar, 2020). The iron and steel plant produces huge amount of wastewater from various unit operations. The pollutants released at various stages of a typical steel industry

 DOI: 10.1201/9781003366263-2

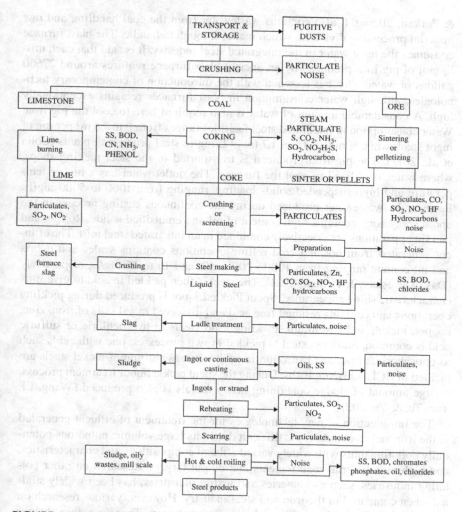

FIGURE 2.1 Pollutants released at various stages of a typical steelmaking operation. (Reproduced with permission from Jorgensen, 2008, Copyright © Elsevier.)

are shown in Figure 2.1. Coke ovens consume the most water and produce the most effluent in an integrated steel plant (Das et al., 2018; Samanta et al., 2021). Quenching, washing, cooling, and isolation of coke industry byproducts are all hydro-intensive procedures. Concentrations of very toxic substances such as oil and grease, cyanides, phenol, thiocyanate are produced during these operations. The tar and liquor plant also consumes huge amount of water to manage water circulation between the coke oven batteries and the by-product recovery area. The wastewater generated is an extremely hazardous ammoniacal liquor that contains ammonia, iron, cyanide, and phenol (Deepti, Sinha, Biswa, Sarkar, Bora

& Purkait, 2020a). It also contains wastewater from the coal handling and raw material processing divisions, which makes it tough to handle. The blast furnace consumes the most water in the integrated steel industry. It is said that each milligram of pig iron produced by an ancient blast furnace requires around 77600 gallons of water. This has lessened with the introduction of contemporary technologies, although water consumption in blast furnaces remains exceptionally high. A considerable amount of water is also required here to cool the pig iron. Water consumption in integrated steel facilities ranges from 0 to 7.6 m^3 per mg of ingot steel, while it ranges from 1.7 to 49 L/mg in steel processing plants (Colla et al., 2017). The pig iron produced is transported to the steel melting shops, where water is utilized to cool the furnace. The outlet water has a higher temperature and high suspended solids loading, ranging from 1000 to 5000 mg/L. Further, the wastewater produced during the continuous casting process in the apron spray zone contains high content of oil and emulsified solids. Rolling and pickling operations are effectively combined in an integrated steel mill. The effluent generated from hot and cold rolling operations contains scales and debris and oil content ranging from 100 to 200 mg/L and 10 to 25 mg/L, respectively (Das, Anweshan, & Purkait, 2021). The steel is then pickled in acid to give it its characteristic shiny appearance. Spent Pickle Liquor is produced during pickling operations and contains residual free acid and dissolved metal salts of iron, zinc, copper, nickel, and chromium. Pickling mild steels in hydrochloric or sulfuric acid is common. Stainless steel is pickled in two processes: one with acids such as phosphoric, nitric, and hydrofluoric acid, whereas chromium-nickel steels are pickled in HCl and HNO_3 baths. During the spent pickle liquor treatment process, a large amount of sludge containing heavy metals is also produced (Wang, Li, Yan, Tu, & Yu, 2021).

The introduction of new technologies for the treatment of effluent generated in the iron and steel industry is difficult due to its large volume, numerous potentially polluting units with widely varied effluent compositions and characteristics, and high raw material requirements and output targets. Research on other polluting industries, such as tanneries and textile industries, has been widely studied when compared to the iron and steel industry. However, various research on the disposal of coke oven effluent has been conducted. Therefore, this chapter focused on the iron and steel industry, covering the majority of the processes and rigorously analyzing the origin, composition, and characteristics of wastewater in each unit.

2.2 CLASSIFICATION OF IRON AND STEEL INDUSTRY WASTEWATER

2.2.1 IRONMAKING WASTEWATER

A blast furnace is a basic unit for producing iron from iron ore, and it produces blast furnace gas as a valuable byproduct, which can be used to fuel some auxiliary stoves and boilers. This gas contains dust, which is partly cleaned by dry dust catchers and subsequently by one or two-stage water scrubbers. The scrubbers used typically work in an extended recycle mode, removing the majority of suspended solids and controlling dissolved solids through the discharge of a blowdown, which must be treated before release to recipient water bodies. The flow of the blowdown and its parameters vary greatly between steel mills due to a variety of variables. The increased water recycling in the gas cleaning system increases the potential for scaling. Opposing viewpoints have been reported on the problem, and several methods to scale control have been proposed, including CO_2 retention in the recycling, acid injection, and adding scaling inhibitors such as organic phosphonate compounds (Ganczarczyk & Lee, 1986)

2.2.1.1 Origin and Composition of Ironmaking Wastewater

A blast of hot air is blown into the blast furnace (BF) during metal manufacturing. The flue gas emerges through the top of BF after experiencing the requisite chemical processes. This gas comprises dust, flammable gas, a high concentration of suspended solids, and high temperature. Therefore, the flue gas is cleaned using a sophisticated method. It is first sent through a dust collector to remove coarser particles and then to a wet cleaning system. At this stage, water gets in contact with air, and practically all of the suspended particles (more than 99%) get separated from the gas and dissolve in water. Polluted water from the system contains a high quantity of suspended particles in a range of 500–10000 mg/l, which is usually sent to cleaning basins, where the suspended particles are precipitated. The entire volume of water gets collected in a reservoir, where suspended materials are sorted by sedimentation. Even after these treatments, water might still have high concentrations of particulates and chloride, sulfate and other ions (Mukherjee, Mondal, Sinha, Sarkar, & De, 2016).

2.2.1.2 Characteristics of Ironmaking Wastewater

The gas from the blast furnace is usually treated to remove dust particles by water spraying. This wastewater mainly contains various components such as phenol, metal ions, cyanides, sulfur, dust, slag, and ore particles (Jorgensen, 2008). The composition of wastewater from wet cleaning of blast furnace gas is shown in Table 2.1.

2.2.2 COKE OVEN WASTEWATER

Metallurgical coke is the second most important element in the steelmaking process after iron. Coal is frequently burned in steel mills to provide power to the iron and steel industry. Coal is turned into metallurgical coke, which is used in blast furnaces in steelmaking enterprises around the world. Millions of tons of coke are generated to supply the demand for metallurgical coke in the face of ever-increasing steel demand. Coal carbonization produces not just coke but also gas and other byproducts like benzene, toluene, naphthalene, and coal-tar compounds, which are used to make various synthetic colors, pharmaceuticals, and high explosives (Chakrabortty, Nayak, Pal, Kumar, & Chakraborty, 2020). As

TABLE 2.1
Composition of Ironmaking- Blast Furnace Wastewater.
(Reproduced with permission from Jorgensen, 2008,
Copyright © Elsevier.)

Parameters	Unit	Range of Values
pH	mg/L	7–9
Total Suspended Solids	mg/L	330–670
Volatile Suspended Solids	mg/L	200
Total Dissolved Solids	mg/L	800–4000
Volatile Solids in the Dry Residue	mg/L	100–320
Permanganate	mg/L	40–150
Cyanide	mg/L	0.6–1.3
Thiocyanates	mg/L	0–17
Phenol Traces Sludge	%	0.5–1.2
Iron	mg/L	140–1180

huge amount of freshwater is consumed for coke making, almost half of the water consumed return as very hazardous wastewater, with a variety of hazardous and toxic contaminants such as oil and grease, ammonia, phenol, thiocyanates and cyanides in varying concentrations depending on the type of coal and the operating conditions in the coke ovens. In many situations, quenching is carried out by wastewater to lower the load of wastewater, resulting in emissions of highly carcinogenic organic compounds into the environment through air, followed by surface water and groundwater. The surface water and groundwater gets contaminated immediately because of the fact that phenolic components are highly soluble in water, making it viable to travel throughout various aquatic environment (Kumar & Pal, 2015).

2.2.2.1 Origin and Composition of Coke Oven Wastewater

Coke is formed through the destructive distillation of coal. During the process, crushed coal is charged to the oven, where coal gas is steadily created and pulled. The generated gas is extracted and routed through various coolers to collect the coal tar. Water is also used to quench hot coke and for coal gas purification, which is formed as powerful ammonia liquor containing high quantities of harmful chemicals like phenol, cyanide, and ammonia. Apart from these chemicals, components such as oil, tar, and grease are mostly discharged, which prevents oxygen in the air from being exposed to fire. Coke oven plants are technologically complicated facilities including coking plants that prepare and supply coal blend from coal, draw coal from coal towers, heat coke ovens, coke pushing, quenching, collection, and cleaning of coke oven gas and coal byproduct processing. Phenolic wastewater and waste generated during gas cooling and washing are the main categories of wastes that are generated in a coke oven unit with wet coke quenching (Dhoble & Ahmed, 2019).

2.2.2.2   Characteristics of Coke Oven Wastewater

The gas released during coal combustion comprises important compounds that can be recovered as a byproduct in the plant, namely, ammonia, phenol, naphthalene, toluene, benzene, and xylene. Coal quality and coke oven operation technology are the two main factors that influence the variation of the composition of effluent characteristics. The gases are rinsed with water first, followed by the production of weak ammonia liquor during the recovery process. The majority of water required in coke plants is for the cooling process, and the wastewater generated during the operation is significant and tough to handle. Phenolic effluent is created during the coking of coal and the extraction of coal derivatives (Mishra, Paul, & Jena, 2021). It has an extremely complex chemical composition. The primary compounds found in coke wastewater are oil, tar, ammonia, phenols, sulfide, cyanide, and thiocyanate. Phenolic chemicals and cyanide are the most important pollutants to be concerned about due to their solubility in water (Das, Mondal, et al., 2021; Mondal & Purkait, 2017; Mondal, Samanta, Meghnani, & Purkait, 2019).

The wastewater from coke plants contains a large volume of suspended particles as well as suspended coke breeze. Surface water was observed to be turning black at the discharge point due to the accumulation of coke breezes. As a result, in addition to surface water contamination and siltation on the riverbanks, a huge amount of coke breeze is wasted every day. The toxicity of phenol is increased by various environmental factors, such as the quantity of dissolved oxygen decreases, rise in salinity, and temperature increases. The high content of BOD, COD, and phenol levels in wastewater cause acute surface water pollution (Mishra, Paul, & Jena, 2018). Treatment of coke oven effluent and regulations are required to manage wastewater from coke oven plants. The composition of coke oven plant effluent is shown in Table 2.2.

2.2.3   COLD ROLL MILL WASTEWATER

Rolling is one of the oldest procedures for lowering the cross-section of a metal sheet, alongside forging, casting, extrusion, and other metal shaping processes. Products obtained after cold rolling are among the top-rated products in the steel industry. According to the World Steel Association, the global average annual export of cold-rolled products was 39.4 million tons (2014–2019), accounting for 8.8% of total steel exports. Picking is the initial operation in the cold rolling mill, and it involves using hydrochloric, sulfuric, or nitric acid to remove the oxide deposit that accumulates during hot rolling. Following a degreasing stage, the material is cold-reduced by compression between rollers, and its metallurgical qualities may be adjusted by annealing. A final rolling stage or skin is applied to the product to improve surface hardness (Nagarajappa, Rao, Tripathy, & Vanitha Manager, 2018). As a result, several million tons per year of cold rolling wastewater, comprising emulsion wastewater, pickling and chromium-containing wastewater, and tempering lubricant wastewater, is generated. The hardest to handle is cold rolling emulsion wastewater, which accounts for the maximum of total cold

TABLE 2.2
Composition of Coke Plant Effluent

Parameters	Unit	Min. Value	Max. Value	Average	Tolerance Limit, IS: 2490 Effluent
pH	–	8.2	8.5	8.35	5.5–9
Temperature	°C	30	33	31.5	40
BOD5	mg/L	63.92	82.34	73.13	30
COD	mg/L	436.7	643.8	540.25	250
Dissolved Solids	mg/L	943.2	1128.4	1035.8	2100
Phenol	mg/L	170.27	192.34	181.30	–
Cyanide	mg/L	25.6	30.2	27.9	0.2
Total Hardness	mg/L	428.7	440.2	434.45	–
Mercury	μg/L	28.9	31.08	30.35	10
Lead	μg/L	0.031	0.065	0.048	100
Arsenic	μg/L	10.65	15.74	13.19	200
Iron	μg/L	0.743	1.338	1.04	–
Zinc	μg/L	0.013	0.028	0.020	5000
Nickel	μg/L	0.048	0.053	0.050	3000

rolling wastewater containing components such as grease, surfactant, detergent, and dilute alkali. Cold rolling emulsion wastewater, being the greatest volume of wastewater, has a great potential to affect the environment and human health if not properly treated (Yun-Hua & Fu-Xing, 2010).

2.2.3.1   Origin and Composition of Cold Roll Mill Wastewater

With numerous sub-processes such as pickling, rolling, annealing, and tempering, the cold rolling mill (CRM) process imparts properties such as reduction, hardness, thickness, and preferred finishing on steel. A significant amount of oily wastewater is produced during the process; the most challenging is emulsion wastewater from the cold rolling process. Cold rolling mill wastewater is a leftover from the steelmaking process with a strong carbonation potential due to its alkaline property. Mainly, CRM effluent comprises acid, phenol, iron and oil, and grease, different salts of copper, chromium, nickel, and zinc based on the composition of the steel. The effluent poses a serious threat to the environment if it is discharged into the water bodies (Das, Anweshan et al., 2021).

2.2.3.2   Characteristics of Cold Roll Mill Wastewater

Rolling mills generate wastewater through the process of cooling bearings and shafts. Water usage is around 10 m/ton, but recirculation can lower this quantity of water by a ratio of 2:7. The composition of wastewater varies greatly. Table 2.3 shows the average composition of cold roll mill wastewater (Jorgensen, 2008)

TABLE 2.3

Composition of Cold Roll Mill Wastewater.

(Reproduced with permission from Jorgensen, 2008, Copyright © Elsevier)

Parameters	Unit	Range of Values
Alkalinity	mg/L	3–4
Total Suspended Solids	mg/L	1000–1500
Volatile Suspended Solids	mg/L	10–100
Sulphate	mg/L	400–500
Total Dissolved Solids	mg/L	10–150
Volatile Solids in the Dry Residue	mg/L	100–150
Lubricating Oil	mg/L	2–50

2.2.4 STEELMAKING WASTEWATER

Steelmaking is an important process in the iron and steel industries. The basic oxygen furnace (BOF), the electric arc furnace (EAF) methods, and the ladle furnace (LF) refining process of steelmaking are the fundamental processes of steelmaking. Unlike other operations, the steelmaking process does not generate much wastewater, but it does generate a lot of solid waste in the form of slags. Nevertheless, wastewater is produced as a result of slag crushing.

2.2.4.1 Origin and Composition of Steelmaking Process Waste

Basic oxygen furnaces (BOF), which are situated in an integrated steel plant with a blast furnace (BF), are supplied molten iron from the BF and steel scraps. The charge fed to the furnace typically comprises steel scrap (10–20%) and molten iron (80–90%). Steel scraps play a substantial function in cooling the furnace and keeping the temperature between 160°C and 165°C for the required chemical reactions to occur. During the oxygen blowing cycles, the furnace is also charged with fluxing chemicals, such as lime or dolomite, to remove undesirable chemical components from the melt. The byproduct obtained after these operations is called slag, which floats on top of the molten steel. The slag produced during the steelmaking cycle is treated, and the finished product is known as the basic oxygen furnace (BOF) slag (Deepti, Sinha, Biswa, Sarkar, Bora & Purkait, 2020b). The chemical reactions that occur during impurity removal govern the chemical composition of slag. Similarly, electric-arc furnaces provide the heat required to melt recyclable steel scrap and turn it into high-quality steel using high-power electric arcs rather than gaseous fuels. During the process of refining, oxygen is added to the molten steel, resulting in the oxidization of iron, silicon, phosphorous, carbon, and aluminum that are present in the heated metal. Slag is formed when these oxidized components interact with lime (CaO).

After the initial steelmaking operations are completed, steel formed by the initial processes can be refined to achieve the necessary composition. Secondary steelmaking operations refer to these refining processes. Refining procedures are very common in manufacturing excellent-quality steels (Huaiwei & Xin, 2011). The utmost essential roles of secondary refining operations are final desulfurization, oxygen, nitrogen, and hydrogen degassing, impurity removal, and final decarburization. Depending on the required steel quality, molten steel produced in the primary processes goes through any or all of the refining procedures listed above. The majority of integrated plants contain secondary metallurgical refining units. Usually, the bottom of the ladle furnace has a conduit through which argon gas is pumped to stir and homogenize the liquid steel in the furnace. The sulfur concentration in steel can be reduced to 0.0002% by injecting desulfurizing chemicals through a lance. Adding silicon and aluminum during deoxidation produces silica and alumina that ultimately reach the refining slag (Yildirim & Prezzi, 2011).

As previously mentioned, basic oxygen furnace and electric arc furnace slags are both generated during basic steelmaking activities. As a result, the chemical and mineralogical compositions of these slags are almost similar in general. The two principal chemical constituents of both slags are calcium oxide and iron oxide. The chemical composition of ladle slag differs from that of BOF and EAF slags.

2.2.4.2 Characteristics of Steelmaking Wastes

The primary chemical elements of basic oxygen furnace slag are calcium oxide, iron oxide, and silicon dioxide. The chemical composition of the BOF slag contains oxidized iron because some portion of the iron in the hot metal is not retrieved in the course of the conversion of molten iron into steel. The percentage of iron oxide in BOF slag can be as high as 38% depending on furnace performance; this is the quantity of oxidized iron that cannot be recovered during the conversion of molten iron into steel. BOF slag mostly consists of silica, aluminum oxide, magnesium oxide, and free lime. The calcium oxide level in BOF slag is often very high due to the utilization of lime or dolomitic lime during the iron-to-steel conversion process (Belhadj, Diliberto, & Lecomte, 2012).

The chemical composition of electric arc furnace (EAF) slag is identical to that of blast furnace slag. The steelmaking process in an EAF is simply a recycling of steel scrap. As a result, the chemical composition of slag is heavily influenced by recycled steel's qualities. The major chemical elements of EAF slags might differ significantly when compared to BOF slags. EAF slags typically have ferric oxide, calcium oxide, silica, aluminum oxide, and magnesium oxide, along with other oxidized impurities, in minute concentrations (Motz & Geiseler, 2001).

Only a few studies have been reported on the composition of ladle furnace (LF) slag. Various alloys are fed to the ladle furnace throughout the steel refining process to get the preferred steel grade. As a result, the composition of slag is strongly reliant on the steel grade and varies greatly when compared to BOF and EAF slags. The content of iron oxide (10%) in ladle slags (LS) is substantially lower than that of the other two slags. Further, ladle slags often have aluminum oxide and calcium oxide concentrations (Yildirim & Prezzi, 2011). The composition of various slags from steel making process is shown in Table 2.4.

TABLE 2.4
Composition of BOF, EAF, and Ladle Slag

Constituents	BOF	EAF	Ladle
SiO_2	12.16	16.1	31.42
Al_2O_3	1.22	7.6	2.95
FeO	26.30	–	–
CaO	47.88	29.5	46.14
MnO	0.28	4.5	1.49
MgO	0.82	5.0	2.13
P_2O_5	3.33	0.6	–
S	0.28	–	–
Na_2O	0.036	–	1.27
K_2O	0.071	–	0.07
Fe_2O_3	–	32.56	0.88
TiO_2	–	0.78	0.79

As discussed before, the steelmaking process does not contribute much to the production of wastewater. However, the wastewater generated is obtained because of slag crushing. The temperature of the wastewater is between 30°C and 40°C. Within three hours, the majority of the suspended solids get settled. The amount of wastewater used during slag crushing varies by plant. In most circumstances, treated wastewater can be reused after purification. The wastewater has a temperature range of 330–500°C. The suspended solid sinks quickly. The suspended debris settles after almost 30 min. The composition of wastewater from slag crushing is shown in Table 2.5 (Jorgensen, 2008).

TABLE 2.5
Composition of Wastewater from Slag Crushing. (Reproduced with permission from Jorgensen, 2008, Copyright © Elsevier)

Parameters	Unit	Range of Values
Alkalinity	mg/L	3–4
Total Suspended Solids	mg/L	500–600
Volatile Suspended Solids	mg/L	30–50
Total Dissolved Solids	mg/L	450–550
Loss of Ignition	mg/L	100–150
Permanganate	mg/L	100–500
Thiocyanates	mg/L	3–4
Phenol Traces Sludge	%	1–2.5
Sulfite	mg/L	10–30
Sulfate	mg/L	100–150

2.3 SUSTAINBILITY IN STEEL INDUSTRIES – A TECHNO-ECONOMIC APPROACH

The water utilized (almost 90%) by the integrated iron and steel industry yields as wastewater, from where valuable chemicals, acids, and metals can be recovered. Strategies such as recovery and reuse should be implemented to improve the sustainability of the steelmaking process by lowering the pollutant burden and accomplishing the process more cost effective. Given the increasing demand for water and its decreasing availability, simple end-of-pipe treatment will not suffice to reduce pollution. Where possible, reusing, recycling, and recovering by-products must become an integrated element of the treatment process (Patwardhan, 2008). Experience has shown that it is possible to attain this goal without incurring significant costs. In many situations, the practice of reusing, recycling, and recovering by-product has resulted in not only meeting the operating costs but also providing the sector with an appealing payback period. Several technologies, including precipitation, membrane process, electrochemical, adsorption, and hybrid process, are widely employed to treat wastewater generated by various steel industry operations. Furthermore, as stated in earlier sections, a substantial quantity of slag is produced from various unit processes of the steel industry. Steel slag is generally made up of calcium, iron, silicon, and aluminum oxides. The Indian steel industry produces approximately 24 million tons of slag per year. Because of the slag's intrinsic physicochemical qualities, it could be utilized as an adsorbent, membrane filter, construction material, bricks, and as a raw material in cement manufacturers (Rawat, Srivastava, Bhatnagar, & Kumar, 2023). The subsequent chapters provide a detailed overview of the available techniques as well as some case studies with respect to wastewater treatment and slag utilization.

2.4 SUMMARY

Integrated iron and steel industries are one of the primary sectors that add a significant share to global economic growth. However, waste generation is massive due to the industry's high production rates. A large amount of water is utilized in the steel industry for various processes such as cooling, descaling, and dust cleansing. The integrated steel plant uses about 28.6 m³/ton of steel produced, while water is discharged in an average of 25.3 m³/ton of steel. Almost 90% of the water used in the steel industry must be treated before it can be reused/discharged into the environment. Units such as mining, pig iron smelting in a blast furnace, Linz-Donawitz converters, and rolling mills generate wastewater in iron and steel plants. All the operations generate wastewater that is highly basic in nature with high concentrations of chlorides, sulfates, calcium bicarbonate, iron, cyanide, fluoride, phenol traces, total and volatile suspended solids, total dissolved solids, cyanide, thiocyanates, and magnesium oxide. Effective treatment technologies are required to increase the quality of wastewater with reference to competence, scalability, and lower energy consumption for steel industrial wastewater treatment. Along with wastewater, an enormous amount of slag and other by-products are also produced. Due to storage restrictions for solid waste and other environmental concerns, recycling such waste materials have become a vital issue for the protection of natural resources. Aside from such environmental issues, several attempts have been made in recent decades to utilize industrial wastes in industrial applications for energy and cost savings, as well as their recycling into usable by-products. The utilization of steel slags as byproducts via recycling is extremely reliant on the slag characteristics and the heat treatment that the molten metal has undergone. Open dumping and landfill operations are the most commonly used methods for disposing of industrial slag, they cause environmental degradation in the form of dust and leachate, in addition to significant economic liability. As a result, extensive research into the use of slag for various beneficial applications is required.

REFERENCES

Belhadj, E., Diliberto, C., & Lecomte, A. (2012). Characterization and activation of Basic Oxygen Furnace slag. *Cement and Concrete Composites*, 34(1), 34–40. https://doi.org/10.1016/j.cemconcomp.2011.08.012

Chakrabortty, S., Nayak, J., Pal, P., Kumar, R., & Chakraborty, P. (2020). Separation of COD, sulphate and chloride from pharmaceutical wastewater using membrane integrated system: Transport modeling towards scale-up. *Journal of Environmental Chemical Engineering*, 8(5), 104275. https://doi.org/10.1016/j.jece.2020.104275

Changmai, M., Das, P. P., Mondal, P., Pasawan, M., Sinha, A., Biswas, P., Sarkar, S., & Purkait, M. K. (2022). Hybrid electrocoagulation–microfiltration technique for treatment of nanofiltration rejected steel industry effluent. *International Journal of Environmental Analytical Chemistry*, 102(1), 62–83.

Colla, V., Branca, T. A., Rosito, F., Lucca, C., Vivas, B. P., & Delmiro, V. M. (2016). Sustainable reverse osmosis application for wastewater treatment in the steel industry. *Journal of Cleaner Production*, 130, 103–115. https://doi.org/10.1016/j.jclepro.2015.09.025

Colla, V., Matino, I., Branca, T. A., Fornai, B., Romaniello, L., & Rosito, F. (2017). Efficient use of water resources in the steel industry. *Water (Switzerland)*, 9(11), 1–15. https://doi.org/10.3390/w9110874

Das, P., Mondal, G. C., Singh, S., Singh, A. K., Prasad, B., & Singh, K. K. (2018). Effluent treatment technologies in the iron and steel industry - A state of the art review. *Water Environment Research*, 90(5), 395–408. https://doi.org/10.2175/1061430 17x15131012152951

Das, P. P., Anweshan, A., & Purkait, M. K. (2021). Treatment of cold rolling mill (CRM) effluent of steel industry. *Separation and Purification Technology*, 274(April), 119083. https://doi.org/10.1016/j.seppur.2021.119083

Das, P. P., Mondal, P., Anweshan, Sinha, A., Biswas, P., Sarkar, S., & Purkait, M. K. (2021). Treatment of steel plant generated biological oxidation treated (BOT) wastewater by hybrid process. *Separation and Purification Technology*, 258(P1), 118013. https://doi.org/10.1016/j.seppur.2020.118013

Deepti., Sinha, A., Biswas, P., Sarkar, S., Bora, U., & Purkait, M. K. (2020a). Separation of chloride and sulphate ions from nanofiltration rejected wastewater of steel industry. *Journal of Water Process Engineering*, 33(December 2019), 101108. https://doi.org/10.1016/j.jwpe.2019.101108

Deepti., Sinha, A., Biswas, P., Sarkar, S., Bora, U., & Purkait, M. K. (2020b). Utilization of LD slag from steel industry for the preparation of MF membrane. *Journal of Environmental Management*, *259*(September 2019), 110060. https://doi.org/10. 1016/j.jenvman.2019.110060

Dhoble, Y. N. & Ahmed, S. (2019). Treatment of wastewater generated from coke oven by adsorption on steelmaking slag and its effect on cementitious properties. *Current Science*, *116*(8), 1346–1355. https://doi.org/10.18520/cs/v116/i8/1346-1355

Ganczarczyk, J. J., & Lee, G. S. 1986). Physico-chemical treatment Of effluents From integrated steel mills. *Studies in Environmental Science*, *29*, 357–367.

Huaiwei, Z. & Xin, H. (2011). An overview for the utilization of wastes from stainless steel industries. *Resources, Conservation and Recycling*, *55*(8), 745–754. https://doi. org/10.1016/j.resconrec.2011.03.005

Jorgensen, S. E. (2008). Waste water from the iron and steel industry and mining. *Studies in Environmental Science*, *5*(), 217–227.

Khunte, M. (2018). Process waste generation and utilization in steel industry, *3*(1), 1–5. https://doi.org/10.11648/j.ijimse.20180301.11

Kumar, R., & Pal, P. (2015). A novel forward osmosis-nano filtration integrated system for coke-oven wastewater reclamation. *Chemical Engineering Research and Design*, *100*, 542–553. https://doi.org/10.1016/j.cherd.2015.05.012

Mishra, L., Paul, K. K., & Jena, S. (2018). Characterization of coke oven wastewater. *IOP Conference Series: Earth and Environmental Science*, *167*(1). https://doi. org/10.1088/1755-1315/167/1/012011

Mishra, L., Paul, K. K., & Jena, S. (2021). Coke wastewater treatment methods: Mini review. *Journal of the Indian Chemical Society*, *98*(10), 100133. https://doi. org/10.1016/j.jics.2021.100133

Mondal, P., & Purkait, M.K. (2017). Green synthesized iron nanoparticle-embedded pH-responsive PVDF-co-HFP membranes: Optimization study for NPs preparation and nitrobenzene reduction. *Separation Science and Technology*, *52*(14), 2338–2355.

Mondal, P., Samanta, N.S., Meghnani, V., & Purkait, M.K. (2019). Selective glucose permeability in presence of various salts through tunable pore size of pH responsive PVDF-co-HFP membrane. *Separation and Purification Technology*, *221*, 249–260.

Motz, H., & Geiseler, J. (2001). Products of steel slags an opportunity to save natural resources. *Waste Management*, *21*(3), 285–293. https://doi.org/10.1016/S0956-053X(00)00102-1

Mukherjee, R., Mondal, M., Sinha, A., Sarkar, S., & De, S. (2016). Application of nano-filtration membrane for treatment of chloride rich steel plant effluent. *Journal of Environmental Chemical Engineering*, *4*(1), 1–9. https://doi.org/10.1016/j. jece.2015.10.038

Nagarajappa, D. P., Rao, R. T. S., Tripathy, A., & Vanitha Manager, C. H. (2018). Sustainable water recovery from cold roll mill of integrated steel plant of JSW steel limited in Torangallu, Karnataka. *GRD Journal for Engineering*, *3*. Retrieved from www.grdjournals.com

Patwardhan, A. (2008). Industrial Wastewater Treatment, PHI Learning Pvt. Ltd., Volume 6.

Purkait, M. K., Sinha, M. K., Mondal, P., & Singh, R. (2018a). Interface Science and Technology, Elsevier.

Purkait, M.K., Sinha, M.K., Mondal, P., & Singh, R. (2018b). Chapter 2 – pH-responsive membranes. *Interface Science and Technology*, *25*, 39–66.

Purkait, M.K., Sinha, M.K., Mondal, P., & Singh, R. (2018c). Chapter 4 - Photoresponsive Membranes, Editor(s): Mihir Kumar Purkait, Manish Kumar Sinha, Piyal Mondal, Randeep Singh, Interface Science and Technology, Elsevier, Volume 25, Pages 115–144, ISBN 9780128139615.

Purkait, M.K., Mondal, P., & Chang, C.-T. (2019). Treatment of Industrial Effluents: Case Studies, CRC Press, 1st Edition, ISBN 9781138393417.

Purkait, M.K., Singh, R., Mondal, P., & Haldar, D. (2020). Thermal Induced Membrane Separation Processes, Elsevier, ISBN 9780128188019

Rawat, A., Srivastava, A., Bhatnagar, A., & Kumar, A. (2023). Technological advancements for the treatment of steel industry wastewater : Effluent management and sustainable treatment strategies. *Journal of Cleaner Production*, *383*(August 2022), 135382. https://doi.org/10.1016/j.jclepro.2022.135382

Samanta, N. S., Banerjee, S., Mondal, P., Anweshan, Bora, U., & Purkait, M.K. (2021). Preparation and characterization of zeolite from waste Linz-Donawitz (LD) process slag of steel industry for removal of Fe3+ from drinking water. *Advanced Powder Technology*, *32*(9), 3372–3387.

Wang, X., Li, X., Yan, X., Tu, C., & Yu, Z. (2021). Environmental risks for application of iron and steel slags in soils in China: A review. *Pedosphere*, *31*(1), 28–42. https://doi.org/10.1016/S1002-0160(20)60058-3

Yildirim, I. Z. & Prezzi, M. (2011). Chemical, mineralogical, and morphological properties of steel slag. *Advances in Civil Engineering*, *2011*. https://doi.org/10.1155/2011/463638

Yun-Hua, Z. & Fu-Xing, G. A. N. (2010). New integrated processes for treating cold-rolling mill emulsion wastewater, *17*(6), 32–35. https://doi.org/10.1016/S1006-706X(10)60110-0

3 Environmental Impacts of Wastewater Generated in Steel Industry

3.1 INTRODUCTION

The high intensity of material and energy consumption involved in the iron and steel industry needs immediate attention since a tremendous amount of pollution is escalated in the form of CO_2 emission and particulate matter emission (Li, Sun, Zhao, & Cai, 2019). This industry involves intensive water utilization and is also a major source of wastewater discharge. For instance, an integrated steel plant utilizes an average of 28.6 m^3 of water in order to process per ton of crude steel according to World Steel Association (2018a). Moreover, data also revealed that the amount of freshwater averages 3.3 m^3, and about 20GJ of energy is consumed per ton of crude steel, which generates average CO_2 emission of 1.9 tons (World Steel Association, 2018b). The calculation of Wang, Wang, Hertwich, and Liu (2017) indicated that 4.1 m^3 of the water use was related to the water-energy-emission network per ton of crude steel production, accounting for around 66% of the total water consumption in the steel production processes. The data reported shows the intensive use of fresh water in the steel industry; hence, reducing fresh water consumption along with wastewater discharge is of great importance.

Massive land areas covering several square kilometers are being utilized by integrated steel plants for handling raw materials and setting up different sections such as coke oven plants, sintering plants, steel smelting workshops, blast furnaces, rolling mills, oxygen plants, and others. Complex organic compounds mainly persist in the waste stream generated from steel plants, such as polycyclic aromatic hydrocarbons (PAH), phenols, cyanide, ammonia, benzene toluene xylene (BTX), and cresols (Purkait, Sinha, Mondal, & Singh, 2018a, b, c). Such substances are highly toxic and hazardous in nature, and their treatment is quite challenging. The entire pollution control of the steel plant industry is divided into primary, secondary, and tertiary treatment. Initially, physical separation steps are involved, such as screening, gravity settling, and removal of viscous materials to minimize pollutant amount, after which coagulation-flocculation follows (Purkait, Mondal, & Chang, 2019; Purkait, Singh, Mondal, & Haldar, 2020). Followed by primary treatment, secondary and tertiary treatments such as hydrogen peroxide oxidation, photo-Fenton oxidation, ultraviolet photolysis, electrochemical oxidation, and ozonation are chosen for simultaneous disinfection and minimizing total dissolved solids (TDS).

DOI: 10.1201/9781003366263-3

An integrated methodology was developed by Mahjouri et al. (2017) for Iran's steel industry in order to determine the most appropriate wastewater treatment technology. Researchers have also analyzed by modeling and simulation the scope of implementation for reverse osmosis and ultrafiltration process (Colla et al., 2017). Optimization of the water system was performed by Alcamisi et al. (2015) utilizing a pipeline networking perspective. The work includes traditional water pinch technology in which the wastewater treatment models are integrated. The initial assessment of water discharge from steel was paid less attention. Evaluation of water usage in the iron and steel industry is done based on two indices: water consumption per ton of crude steel and freshwater consumption per ton of crude steel (Tian, Zhou, & Lv, 2008). Intake of water from the surface and underground is termed freshwater consumption. The total summation of the freshwater and circulating water is termed water consumption. China is found to contribute about 14% of the total wastewater generated from the iron and steel industry specifically (Guo & Fu, 2010). A quantitative evaluation of water consumption status can be done by using these two indices. Moreover, researchers have also devoted their studies to finding the environmental impact of water quality that is being discharged as wastewater. Kanu and Achi (2011) worked on the water quality assessment and figured out that the wastewater generated from the iron and steel industry contained high suspended solids (SS). Various research work concluded that the iron and steel industry generated effluents that are heavily loaded with severe toxic and hazardous pollutants along with unutilized components, requiring remediation. The study by Wang, Liu, Ye, and Sun (2018) focused on the quantification of various hazardous components, such as suspended solids, chemical oxygen demand, pH, and electrical conductivity of converter dust removal sewage. Ma et al. (2018) concluded that in order to access the use of water in the iron and steel industry and to evaluate pollutants environmental impact, water footprint plays a vital role in such assessment, which includes blue, green, and graywater footprint. Considering the iron and steel industry of eastern China as a case study, Gu et al. (2015) calculated the graywater footprint. Yet, the value of water footprint may be the same for different components in wastewater with the same maximum volume of diluting water to meet the environmental standard.

3.2 EFFLUENT PRODUCTION AND ENVIRONMENTAL HAZARDS

Large volumes of effluent are being discharged from different units of the iron and steel industry. Various pollutants are emitted by a steelmaking process from different operation stages. Among other sections, coke oven plants in an integrated steel factory produce the highest effluent and water consumption (Mondal & Purkait, 2017; Mondal, Samanta, Meghnani, & Purkait, 2019). In the steel industry, various operations utilize water extensively, such as quenching of hot coke, cooling and washing of gases from coke ovens, ammonia still washing, and by-products isolation of coke industry. To produce 100 mg of coke, about 40 m^3 of water is required in a steel industry. During the synthesis of coke in the

steel industry, hazardous toxic substances such as thiocyanate, cyanides, ammonia, phenol, grease, and oil are generated (Changmai et al., 2022; Ghose, 2002;
Samanta et al., 2021). Tar and liquor plants are generally placed next to the coke
oven. It handles the circulation of water between the by-product recovery section and the coke oven battery. The circulated water is ammoniacal liquor which
is considered highly toxic due to the presence of ammonia, iron, cyanide, and
phenol. Such generated water is very hard to handle since it includes coal handling wastewater along with toxic water generated from raw material preparation
sections.

In the steel industry, an old blast furnace consumes around 77600 liters of water
in order to produce per Mg of pig iron. The general discharge standards from the
different units of steel plant wastewater is mentioned in Table 3.1. Over the years,
with the introduction of hybrid technologies, the discharge parameters have been
controlled except for the consumption of large volumes of water. In fact, a huge volume of water is also required to cool the pig iron produced. In the blast furnace of
the integrated steel plant, about 7.6 m^3 of water is consumed per Mg of ingot steel,
whereas about 1.7 to 49 L/Mg of water is consumed for steel processing plants. The
steel-making section then uses water to cool the pig iron produced in the furnace.
The wastewater consists of a high suspended solid loading of 1000–5000 mg/L
along with a relatively higher temperature. During the continuous casting process,

TABLE 3.1
Integrated Iron and Steel Plant: Wastewater Discharge Standards
(Data from Das et al., 2018, Copyright © John Wiley and Sons)

Plant	Parameter	Concentration (mg/L, except pH)
Coke Oven By-Product Plant	pH	6.0–8.0
	Suspended Solids	100
	Phenol	1
	Cyanide	0.2
	BOD, 3 days, 27°C	530
	COD	250
	Ammoniacal Nitrogen	50
	Oil and Grease	10
Sinter Plant, Rolling Mills, and Steel Melting Shop	pH	6.0–9.0
	Suspended Solids	100
	Oil and Grease	10
Blast Furnace	pH	6.0–8.5
	Suspended Solids	50
	Oil and Grease	10
	Cyanide	0.2
	Ammoniacal Nitrogen	50

oil leakages and emulsified solids are further found to contaminate the wastewater produced in the apron spray zone. Rolling and pickling operations are generally found to be combined in an integrated steel industry. Scales and debris loading of 100–200 mg/L are generally associated with hot and cold rolling operations, whereas the oil content is about 10–25 mg/L (Biswas, 2013). The characteristic finishing glossy color is obtained after pickling the steel in acid. Spent pickle liquor (SPL) is obtained after the pickling operation, which contains dissolved metal salts of nickel, iron, copper, chromium, and zinc. The effluent also consists of rinse water, residual free acid, and fume scrubber wastewater.

Hydrochloric acid (HCl) or Sulfuric acid (H_2SO_4) are often utilized for pickling of mild steels. Phosphoric, nitric, and hydrofluoric acids are used for pickling stainless steel in two steps. Moreover, HCl and HNO_3 are used to pickle rust and acid-resistant chromium-nickel steels. In the fume scrubber, if acids other than phosphoric or sulfuric are used, the generated wastewater is added to the waste. A huge amount of sludge is generated during the spent liquor processing treatment. Various pollutants that are being discharged from the main unit of a steel plant are presented in Table 3.2.

Since the steel industry generates a huge amount of obnoxious compound loaded effluent, it has rendered the water bodies and large land areas polluted and unsafe for the environment. In the southern part of India, a city named Salem reported higher fluoride levels because of effluents generated by the Salem Steel Plant (Saha & Chandran, 2001). Similarly, the effluent discharged from the Bhilai Steel plant, near the areas of river Senath and its tributary, has polluted the river causing a steep rise in dissolved and suspended solids, with the inclusion of heavy metals such as mercury and cadmium. This has also resulted in decrease in the pH of the river as well as a reduction in dissolved oxygen content. Due to the highly toxic nature of the released effluent from the Bhilai steel plant abrupt decrease in the fish population was recorded in the river (Satish, Chandra, Sar, & Bhui, 2012).

Soni and Bhaskar (2012) studied the physiochemical property variation in the soil of Hisar, a district of Haryana, due to its exposure to the steel industry solid wastes. The study revealed that high levels of magnesium and calcium were found in the samples collected. Many researchers have concluded in their investigation that ammoniacal liquor produced from coke oven plants has drastically deteriorated the environmental quality. The hazardous effect includes the change in the balance of nitrification and de-nitrification process along with cyanide pollution in water and soil. Ghose (2002) studied the hazardous impact of the effluent discharged from the Jharia Coalfields area of Jharkhand. It was reported that the generated wastewater had high chemical oxygen demand (COD), biological oxygen demand (BOD), and suspended solids and phenols, which causes drastic pollution in nearby water bodies. Igwe and Ukaogo (2015) in his study reported that steel industry effluents contain polycyclic aromatic hydrocarbons (PAH), which are considered carcinogenic in nature and potentially bioaccumulative. Hence from the investigations carried out, it is certain that steel plants are contributing toward a huge environmental crisis, and thus, the treatment technologies need to be improved with advanced treatment methods in order to tackle the present situation.

TABLE 3.2
Effluent Characteristics of Different Steel Plant Units (Data from Das et al., 2018, Copyright © John Wiley and Sons)

Section	pH	Alkaline (mg/L)	TDS (mg/L)	TSS (mg/L)	VSS (mg/L)	Anions (mg/L)	Cations (mg/L)	Phenol (%)	Others (mg/L)
Blast Furnace	7–9	–	–	330–350	≤200	$MnO_4^- = 40\text{-}15$ $CN^- = 0.6\text{-}1.3$ $CNS^- = 0\text{-}17$	$Fe^{2+} = 140\text{-}180$	0.5–1.2	–
Slag Crushing	–	3–4	450–550	500–600	30–50	$MnO_4^- = 100\text{-}500$ $SO_4^{2-} = 100\text{-}150$ $CN^- = NA$ $CNS^- = 3\text{-}4$	–	1.0–2.5	–
Rolling Mills	–	3–4	400–500	1000–1500	10–100	$SO_4^{2-} = 100\text{-}150$	–	–	Lube oil = 2–50
Coke Oven	7.0–8.5	–	800–1200	200–7000	–	$CN^- = 8.2\text{-}21$	$NH_4^+ = 48\text{-}1500$	–	Phenol = 82–123 BOD = 64–94
Pig Iron Cooling	7–8	–	500–2000	500–3500	350	$MnO_4^- = 100\text{-}500$ $SO_4^{2-} = 100\text{-}150$ $Cl^- = 200\text{-}300$	–	–	–
Pickling Unit	1.5–4.5	–	–	500–2000	–	$SO_4^{2-} = 200\text{-}2000$	$Fe^{2+} = 80\text{-}600$ $Ca^{2+} = 50\text{-}200$ $Al^{3+} = 0\text{-}50$	–	–

3.3 MEASURES AND REGULATIONS RECOMMENDED FOR STEEL INDUSTRY WASTEWATER MANAGEMENT

Depending on the geographic location, the amount of water usage varies greatly, and it also depends on the structure of production, the configuration of the plant, the quality of water, equipment, and technology utilized along with the water system and its management. Eliminating lower grade equipment and adopting advanced water-saving technologies are critical for solving steel industry water utilization and wastage problems. In view of the problems related to saving water and emission reduction from the iron and steel industry, certain measures and recommendations need to be adopted and implemented. A few recommendations are mentioned below:

3.3.1 SUSTAINABLE AND SCIENTIFIC MANAGEMENT OF WATER RESOURCES

In regions where water scarcity is a problem, water resource management plays a vital role, and its implementation is necessary in terms of sustainability concerns (Ma et al., 2018). The management of water resources basically deals with the maintenance of water in such a way that it can be stored or maintained over an indefinite time without harming the ecosystem socially or economically. In 2007, the World Steel Association officially launched a project entitled "sustainable water management", keeping in mind the sustainability issues (World Steel Association, 2011). In China, "Jigang Steel", known to be one of the largest steel enterprises, have taken certain steps which resulted in better water management. To observe and organize the water resources online, the enterprise has implemented the initiative of installing digitalized water network systems (Wang, Li, Xiong, & Cang 2011). Therefore, based on the present need of steel industries, it is necessary to focus on sustainable strategies for the management of water that could enable the recycling and reuse of water resources. Such initiatives could help the steel industries achieve a "zero-discharge" level and help the environment to remain eco-friendly.

The main sustainable management ideas for the steel industry include:

a. Superior digitization is required to constantly improve water management systems along with better normalization and standardization.
b. Online water monitoring system should include indexing based on the quality and quantity of water being supplied and discharged along with the equipment's running status.
c. Accuracy and precision of the instruments for managing water systems should be ensured.
d. Along with digitalization, advanced technical transformations should be combined for new sustainable water management development plans.
e. Fixed amount of water usage should be implemented in the steel industry plant for specific unit operations.

3.3.2 STEEL INDUSTRY'S WATER SYSTEM OPTIMIZATION

The optimization of steel industry-generated wastewater has been carried out by many researchers by utilizing substance flow analysis along with keeping in mind the strategy for reducing, reusing, and recycling (3R Principle) (Gao et al., 2011). Optimization mainly aims at enhancing water efficiency, saving it, and reducing the wastewater discharge. In order to implement it, firstly, the quality and quantity of water being distributed along with its consumption at different locations should be sorted out. Secondly, by encouraging water recycling, steel manufacturers can significantly reduce the consumption of water for unit processes. For this, waste-water from the scrubber after treatment, and pH correction may recycled as prac-tice at Arcelor Mittal. Another method to minimize water use is to use the water in cascades until single parameters reach their legal or technical limits. Finally, to avoid substantial increment in energy and other costs, water circulation systems containing large, medium, and small and large cycles should not be used blindly (Wakeel, Chen, Hayat, Alsaedi, & Ahmad, 2016).

3.3.3 DEVELOPMENT AND ADOPTION OF NEW WATER-SAVING TECHNOLOGIES AND ELIMATION OF LOW-LEVEL EQUIPMENT

Blast furnace gas (BFG), coke dry quenching (CDQ), and Linz-Donawitz gas (LDG) dry de-dusting technologies are some of the new technologies for sav-ing water and are introduced and developed in the steel industry. Production of coke using newly proposed CDQ technology reduces air emission and also recovers heat during the on-site quenching process. This technique of coke pro-duction is more environmentally friendly (Liu & Yuan, 2016). It is estimated that through the use of the CDQ technique, nearly 0.43 m^3 of water per ton of coke can be saved without wastewater discharge. Thus, it can be said that CDQ is a technique that saves water and energy and is also environmentally friendly and pollution-free. Converters and blast furnaces produce a large volume of gas, having a large amount of dust. BFG and LDG dry dusting techniques have benefits in high de-dusting efficiency, energy and water-saving, clean environ-mental protection, and emission control. They have a significant dominance in the area of water consumption reduction and energy saving, with 70% realizing sewage zero-emission (Zhou, Zhang, Zhang, & Zhang, 2013). Obsolete and old equipment replacement can also reduce freshwater consumption by the iron and steel industries.

3.3.4 DEVELOPMENT OF UNCONVENTIONAL WATER RESOURCES

In order to reduce the cost of production and control the fresh water shortage problem, unconventional resources of water, such as reclaimed water, seawater, and rainwater, should be focused on development (Cheng, Hu, & Zhao, 2009). In 2005 in China, a development plan was issued by the State oceanic admin-istration for seawater utilization. The steel plant and power plant then started utilizing the seawater after desalination as makeup and cooling water for its different units. In the steel industry, reclaimed water is found to be a reliable and comparatively cheaper source of water. The various utilization of such reclaimed water includes; washing slag, gas washing, de-dusting, and greening (Yi, Jiao, Chen, & Chen, 2011).

3.3.5   Increasing the Rate of Sewage Use and Promotion of Water Treatment Techonoligies

After treatment in steel and iron industries, the wastewater generated can be reutilized in order to minimize freshwater demand. Such reusing initiatives can prove to be beneficial to reduce the demand for water from natural resources. The Chinese state council, considering the importance of water, has published a notice, "The regulation of levying and using discharge fee". Based on this, the Chinese government started collecting discharge fees from steel plants. The production cost of steel industries was raised due to such extra charges. Thus, to minimize the cost of production, reusing and recycling wastewater generated from steel plants after proper treatment was adopted. Such re-utilization was found to enhance the reuse rate of sewage in steel plants (Tong et al., 2018). Thus, wastewater treatment utilizing appropriate technology serves the dual purpose of reducing freshwater demand along with preventing contamination of natural bodies.

3.4   CONCLUSION

For the growth and development of a country, the iron and steel industry plays an important role, since the products obtained from the iron and steel industries are important for society. But from another perspective, the demand for water resources increases with an increase in the production of the iron and steel industry, putting deliberate pressure on natural water sources such as groundwater and surface water. Moreover, various wastewater released from different units of the steel industry consists of harmful and toxic pollutants, adversely affecting the environment.

The main aim regarding the treatment of steel industry effluent should not only focus on achieving the discharge standards but also look upon the reusability of such wastewater. Various treatment technologies designed so far for the steel industry wastewater treatment have aimed to reduce the energy of operation and minimize of effluent volume, which are important from an industrial development perspective as well as efficient circulation of the economy. From an industrial perspective, focusing on life cycle assessment (LCA) is necessary to prioritize resource recovery and plan water footprint reduction.

Since the steel industry wastewater comprises a complex mixture of various toxic and hazardous components, its management is also a hectic and challenging task. This chapter thus focuses on the hazards and life risk assessment of the nature of the steel wastewater generated along with the measures and regulations necessary to make it a sustainable and economically viable.

REFERENCES

Alcamisi, E., Matino, I., Colla, V., Maddaloni, A., Romaniello, L., & Rosito, F. (2015). Process integration solutions for water networks in integrated steel making plants. *Chemical Engineering*, 45, 37–42.

Biswas, J. (2013). Evaluation of various method and efficiencies for treatment of effluent from iron and steel industry - a review. *International Journal of Mechanical Engineering and Robotics Research*, 2, 1–9.

Changmai, M., Das, P. P., Mondal, P., Pasawan, M., Sinha, A., Biswas, P., Sarkar, S., & Purkait, M. K. (2022). Hybrid electrocoagulation–microfiltration technique for treatment of nanofiltration rejected steel industry effluent. *International Journal of Environmental Analytical Chemistry*, *102*(1): 62–83.

Cheng, H., Hu, Y., & Zhao, J. (2009). Meeting China's water shortage crisis: current practices and challenges. *Environmental Science & Technology*, *43*(2), 240–244.

Colla, V., Matino, I., Branca, T. A., Fornai, B., Romaniello, L., & Rosito, F. (2017). Efficient use of water resources in the steel industry. *Water*, *9*(11), 874.

Das, P., Mondal, G. C., Singh, S., Singh, A.K., Prasad, B., & Singh, K. K. (2018). Effluent treatment technologies in the iron and steel industry - A state of the art review. *Water Environment Research*, *90*(5), 395–408.

Gao, C., Wang, D., Dong, H., Cai, J., Zhu, W., & Du, T. (2011). Optimization and evaluation of steel industry's water-use system. *Journal of Cleaner Production*, *19*(1), 64–69.

Ghose, M. K. (2002). Physico-chemical treatment of coke plant effluents for control of water pollution in India. *Indian Journal of Chemical Technology*, 9, 54–59.

Gu, Y., Keller, A. A., Yuan, D., Li, Y., Zhang, B., Weng, Q., Zhang, X., Deng, P., Wang, H., & Li, F. (2015). Calculation of water footprint of the iron and steel industry: a case study in Eastern China. *Journal of Cleaner Production*, *92*, 274–281.

Guo, Z. C. & Fu, Z. X. (2010). Current situation of energy consumption and measures taken for energy saving in the iron and steel industry in China. *Energy*, *35*(11), 4356–4360.

Igwe, J. C. & Ukaogo, P. O. (2015). Environmental effects of polycyclic aromatic hydrocarbons. *Journal of Natural Sciences Research*, 5, 117–132.

Kanu, I., & Achi, O. K. (2011). Industrial effluents and their impact on water quality of receiving rivers in Nigeria. *Journal of Applied Technology in Environmental*, *1*(1), 75–86.

Li, X., Sun, W., Zhao, L., & Cai, J. (2019). Emission characterization of particulate matter in the ironmaking process. *Environmental Technology*. *40*(3), 282–292.

Liu, X., & Yuan, Z. (2016). Life cycle environmental performance of by-product coke production in China. *Journal of Cleaner Production*, *112*, 1292–1301.

Ma, X., Ye, L., Qi, C., Yang, D., Shen, X., & Hong, J. (2018). Life cycle assessment and water footprint evaluation of crude steel production: A case study in China. *The Journal of Environmental Management*, *224*, 10–18.

Mahjouri, M., Ishak, M. B., Torabian, A., Manaf, L. A., Halimoon, N., & Ghoddusi, J. (2017). Optimal selection of Iron and Steel wastewater treatment technology using integrated multi-criteria decision-making techniques and fuzzy logic. *Process Safety and Environmental Protection*, *107*, 54–68.

Mondal, P., & Purkait, M. K. (2017). Green synthesized iron nanoparticle-embedded pH-responsive PVDF-co-HFP membranes: Optimization study for NPs preparation and nitrobenzene reduction. *Separation Science and Technology*, *52*(14), 2338–2355.

Mondal, P., Samanta, N. S., Meghnani, V., & Purkait, M. K. (2019). Selective glucose permeability in presence of various salts through tunable pore size of pH responsive PVDF-co-HFP membrane. *Purification Technology*, *221*, 249–260.

Purkait, M. K., Sinha, M. K., Mondal, P., & Singh, R. (2018a). Interface Science and Technology, Elsevier, Pages 115–144.

Purkait, M. K., Sinha, M. K., Mondal, P., & Singh, R. (2018b). Chapter 2 - pH-Responsive Membranes, Singh, Interface Science and Technology, Elsevier, Volume 25, Pages 39–66, ISBN 9780128139615.

Purkait, M. K., Sinha, M. K., Mondal, P., & Singh, R. (2018c). Chapter 4 - Photoresponsive Membranes, Editor(s): Mihir Kumar Purkait, Manish Kumar Sinha, Piyal Mondal, Randeep Singh, Interface Science and Technology, Elsevier, Volume 25, Pages 115–144, ISBN 9780128139615.

Purkait, M. K., Mondal, P., & Chang, C.-T. (2019). *Treatment of Industrial Effluents: Case Studies*, CRC Press, 1st Edition, ISBN: 9781138393417.

Purkait, M. K., Singh, R., Mondal, P., & Haldar, D. (2020). Thermal Induced Membrane Separation Processes, Elsevier, ISBN: 9780128188019.

Saha, S., & Chandran, T. J. (2001). Removal of Fluoride from the Salem steel plant effluent. *Indian Journal of Environmental Protection*, *21*, 627–630.

Samanta, N. S., Banerjee, S., Mondal, P., Anweshan, Bora, U., & Purkait, M. K. (2021). Preparation and characterization of zeolite from waste Linz-Donawitz (LD) process slag of steel industry for removal of Fe^{3+} from drinking water. *Advanced Powder Technology*, *32*(9), 3372–3387.

Satish, S., Chandra, H., Sar, S. K., & Bhui, A. K. (2012). Environmental sinks of heavy metals: Investigations on the effect of steel industry effluent in the Urbanised Location. *International Journal of Advanced Engineering Research*, *1*, 235–239.

Soni, R., & Bhaskar, R. (2012). Impact of steel industry waste on physico-chemical property of soil. *International Journal of Environmental Science*, *2*, 1144–1153.

Tian, J. R., Zhou, P. J., & Lv, B. (2008). A process integration approach to industrial water conservation: a case study for a Chinese steel plant. *Journal of Environmental Management*, *86*(4), 682–687.

Tong, Y., Zhang, Q., Cai, J., Gao, C., Wang, L., & Li, P. (2018). Water consumption and wastewater discharge in China's steel industry. *Iron Making Steel Making*, *45*(10), 868–877.

Wakeel, M., Chen, B., Hayat, T., Alsaedi, A., & Ahmad, B. (2016). Energy consumption for water use cycles in different countries: A review. *Applied Energy*, *178*, 868–885.

Wang, C., Wang, R., Hertwich, E., & Liu, Y. (2017). A technology-based analysis of the water-energy-emission nexus of China's steel industry. *Resources, Conservation, and Recycling*, *124*, 116–128.

Wang, J., Li, S., Xiong, G., & Cang, D. (2011). Application of digital technologies about water network in steel industry. *Resources, Conservation, and Recycling*, *55*(8), 755–759.

Wang, L., Liu, C., Ye, Z., and Sun, W. (2018). Pretreatment of metallurgical sewage via vacuum distillation driven by low-temperature exhausted gas from steel plants. *Journal of Iron and Steel Research International*, *25*(3), 291–297.

World Steel Association. (2011). Water management in the steel industry. https://worldsteel.org/publications/bookshop/water/. Accessed Dec 2022

World Steel Association. (2018a). Water management in the steel industry. https://www.worldsteel.org/publications/position-papers/water-management.html.

World Steel Association. (2018b). Steel's contribution to a low carbon future and climate resilient societies. https://www.worldsteel.org/en/dam/jcr:66fed386-fd0b-485eaa23-b8a5e7533435/Position_paper_climate_2018.pdf

Yi, L., Jiao, W., Chen, X., & Chen, W. (2011). An overview of reclaimed water reuse in China. *Journal of Environmental*, *23*(10), 1585–1593.

Zhou, H., Zhang, F. M., Zhang, D. G., & Zhang, L. Y. (2013). Study on BOF gas dedusting technology at contemporary steel plant. *Advanced Materials Research*, *610*, 1422–1425.

4 Treatment Techniques for Steel Industry Wastewater

4.1 INTRODUCTION

In recent times, heavy industrialization and globalization have paved the way for the iron and steel industry to reach its peak and have attained tremendous growth. In the year 2015, it was reported by the World Steel Association that about 89.6 Mg tons of crude steel was produced worldwide. When the world faces drastic water shortage problems, such statistics of steel production and freshwater usage in the different units of the steel industry are of great concern. Thus, it is very important to provide advancements and progress toward water treatment technologies in order to reduce the usage of fresh water and provide zero effluent discharge in steel industries.

In recent times, membrane technology has evolved to be a better alternative in comparison to the conventional separation processes. Researchers have focused on the energy-saving and eco-friendliness aspects of the membranes and have found that advanced membranes for crystallization purposes can yield better by-product recovery (Minhalma & De Pinho, 2004). Though memory technology has certain disadvantages such as time-dependent flux decline and fouling, research studies have shown that such problems can be overcome with improved antifouling agents and fouling-free cross-flow modules.

Iron and steel industry effluent, due to the presence of numerous toxic pollutants, presents a veritable challenge. Along with the presence of pollutants, the vast volume of water being used, the requirement of large amounts of raw materials, and a high targeted output make introducing of new treatment technology a bit tough in this sector. Various studies have been conducted focusing on the treatment of effluents from tanneries and textile production industries. However, very few have been explored so far regarding the introduction of new treatment technology for the iron and steel industry wastewater. Moreover, few research have focused on the new technologies for the treatment of coke oven effluent in the steel industry sector. This chapter focuses on the various technologies adopted for treating the effluents from steel industry wastewater generated from multiple unit operations. The possible benefits of valuable by-product recovery from the waste streams were explored so as to enhance water recycling and minimize material consumption. The chapter also focuses on the several conventional and advanced processes involved so far in treating steel

38   DOI: 10.1201/9781003366263-4

industry effluent along with a future scope of research views on the present problems faced in the industry.

4.2 CONVENTIONAL TREATMENT TECHNIQUES

4.2.1 ADSORPTION

Adsorption mainly involves adsorbents which are utilized to selectively remove toxic and hazardous compounds from steel industry effluents such as cyanide and phenols from coke oven liquors, removal of heavy metals from casting operations, and eliminating surfactants from pickling liquors (Vazquez, Rodriguez-Iglesias, Maranon, Castrillon, & Alvarez, 2007). Beh, Chuah, Nourouzi, and Choong (2012) utilized furnace slag as an adsorbent to treat the wastewater generated from the steel industry in Malaysia. The results showed a drastic reduction in the concentrations of zinc, iron, copper, and manganese. Moreover, zeolites have been considered to be industrially effective adsorbents for treating and selectively removing targeted compounds. Jami, Rosli, and Amosa (2016) optimized the results of a better adsorbent for removing manganese from steel industry wastewater by comparing naturally available clinoptilolite and commercially synthesized zeolite 3A. The study showed the adsorption variation with contact time and dosage amount, along with the adsorption kinetics. The maximum removal efficiency of 84% was achieved in the study conducted at optimized conditions at an adsorbent dosage of 1 g/mL. The adsorption technique has the main drawback regarding the regeneration of adsorbents after use and thus limits its application only in batch filtration use and not in continuous mode application. The cost of production of adsorbents also sets limitations to the implementation of such technologies. In order to reduce the adsorbent's cost of production, inexpensive waste materials such as seed coats, fruit peels, rice husks, and vegetable peels were utilized. Moreover, bentonite, boron mud haydite, sawdust, and fly ash were utilized by Li, Li, Shao, and Zhang (2011) in order to study the turbidity removal efficiency from steel industry effluent. The turbidity removal efficiency of 94% was attained per hour at a flow rate of 1200 L/s in alkaline pH 12.

Amosa (2015) investigated the utilization of empty fruit bunch (EFB), a waste from the palm oil industry, in the process of wastewater treatment. The EFB is recycled after drying as raw material for preparing powdered activated carbon (PAC) accompanied by carbonization followed by steam pyrolysis. The prepared adsorbent was utilized for treating the effluent of the palm oil mill and was found to achieve a removal efficiency of 95% for manganese and 90% for hydrogen sulfide (H_2S). Due to the high surface area of 886.2 m^2/g, the metal uptake capacity was found to be very high. Amosa et al. (2014) prepared powdered activated carbon from EFB by two methods, utilizing steam and CO_2-based systems, and compared their removal efficiency. The results confirmed that the steam-activated PAC performs better with a higher removal efficiency of 81% for COD, 92% for manganese, and 89% for hydrogen sulfide (H_2S). Activated carbon-based adsorbent processes have also shown tremendous potential in removing organic matter from effluents. Amosa et al. (2014) also concluded that the effect of PAC dosage, contact time, and amount of agitation controls the rate of removal of various pollutants present in the effluent.

Granulated activated carbon filters are the new generation adsorbents which have shown better results in removing pollutants from the wastewater of the iron and steel industry. These materials are synthesized by heating the crude organics in the absence of oxygen. Such a process enhances the surface porosity that helps in selective adsorption and trapping specific chemicals from a complex mixture of industrial effluent. Such adsorption-based processes are robust in nature and are effectively capable of removing microbial and organic pollutants from wastewater (Servais, Billen, & Bouillot, 1994) due to their higher specific surface areas ranging from 1100 to 1250 m^2/g of adsorbent (Choi, Kima, & Kimb, 2008).

For the implementation of advanced technologies in real life, the modeling, simulation, and economic analysis are important aspects which play a vital role in boosting the scale up perspective. The removal rates of eco-toxic micro-pollutants present in wastewater after biological treatment utilizing PAC were evaluated by Nowotny, Epp, von Sonntag, and Fahlenkamp (2007). The study was conducted utilizing simulated conditions considering the presence of probable interfering organic matter that was subsequently validated. In order to propose a fixed bed adsorber design layout, it was observed that the proposed model was appropriate. A convenient in-situ technology for waste remediation was developed by Gao, Shulei, Wang, Gui, and Xu (2016). In this work, directly coking coal was utilized for pre-treatment in order to treat and adsorb impurities from coke oven wastewater. A coking coal dose of 120 g was added, and the mixture was then stirred for 40 min to remove 65% of COD and 34% of phenol from 1 dm^3 of wastewater. The coking coal was then carbonized at a high temperature in order to create coke. According to Gao et al. (2016), the overall cost of recovering water resources might be significantly reduced if effectively implemented.

Nanomaterials are a new type of adsorbent that have gained more attention recently. As adsorbents, carbon nanotubes help to extract both organic and inorganic elements from wastewater (Samanta et al., 2021). Shape, contact areas, average pore diameter and volume, morphology, and functional groups of carbon nanotubes (CNTs) are the factors governing the adsorptive interfaces between them and organic pollutants; whereas, for organic pollutants, hydrophobicity, electron polarizability, polarity, size, functional groups, and environmental conditions (pH, ionic strength) are the dominant factors affecting system performance.

Using carbon nanotubes (CNTs), it has been possible to effectively remove heavy metals like copper and synthetic colors from wastewater (Kabbashi, Karim, Saeed, & Yacoob, 2008). On PAC coated with Fe^{3+} catalysts, researchers produced CNTs in batches using fixed catalyst chemical vapor deposition. The functionalization of the adsorbent was shown to be necessary in order to maximize the reaction time, gas flow rates, and reaction temperature. Both the functionalization processes of sonication with $KMnO_4$ and refluxing with HNO_3 at 140°C were evaluated. Of the two, $KMnO_4$ showed exceptionally high cadmium removal effectiveness from synthetic wastewater (98.35%). However, it is necessary to assess whether such a system is suitable for industrial upscaling. Additionally, research in this area has to concentrate on calculating the techno-economic viability of industrial applications.

4.2.2 COAGULATION AND PRECIPITATION

Coagulation is one of the most widely used traditional effluent treatments for removing contaminants, followed by flocculation and sedimentation employing settling tanks, clarifiers, and clari-flocculation. These technologies are frequently employed to remove emulsified oils found in cold rolling mill residues, to precipitate out the iron and heavy metal residues, and to remove other undesirable elements (Amuda, Amoo, Ipinmoroti, & Ajayi, 2006). The most widely used flocculants include calcium oxide, calcium hydroxide, potash alum, and magnesium salts. These substances work by a variety of methods, including charge neutralization, bridging, electrostatic patch, and so on. Coagulants also help to lower the overall pollutant loading that tertiary treatment processes. Amosa, Jami, Muyibi, Alkhatib, and Jimat (2013) looked at the effectiveness of membrane fouling reduction prior to membrane filtration. Palm oil mill effluent (POME) showed 92.8% COD removal, 99.3% color removal, and 99.9% turbidity removal as a pre-treatment alternative when the wastewater was treated with ferric chloride as a coagulant and polyacrylamide as the flocculant followed by adsorption.

Additionally, chemical coagulation has been used to remove high molecular weight organic compounds, colloids, and particles from effluents (Jami et al., 2016). For the main treatment of wastewater, pre-treatment and chemical coagulation processes are combined, as shown in Figure 4.1. Polymeric coagulants have been developed recently. In order to create composite coagulants with greater

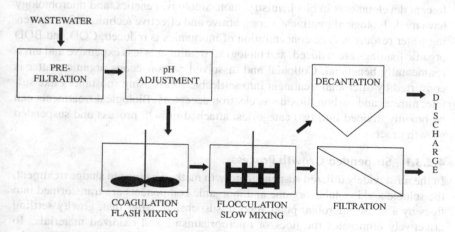

FIGURE 4.1   Coagulation flocculation in primary treatment (Reproduced from Das et al. 2018, Copyright © John Wiley and Sons).

pollutant removal effectiveness, various additives, such as organic molecules and ionic or non-ionic poly-electrolytes, are combined with a predetermined composition. In certain cases, this has resulted in the removal of TDS and total suspended solids (TSS) (Lee, Robinson, & Chong, 2014). However, in industrial settings, the main usage of coagulants, flocculants, and sedimentation is as a pre-treatment step prior to the main treatment, like membrane filtration.

It has long been known that seawater is frequently used to quench coke, producing large concentrations of ions such as fluoride, fluoride, sulfate, and others. The steel industry uses biological oxidation followed by NF to treat coke oven effluent as a solution to the problem. Since the NF-rejected stream contains between 8000 and 10,000 mg/L of salts, disposal is challenging. The chlorides and sulfates from the NF-rejected stream of the steel industry were precipitated using miscible organic solvents, such as diisopropylamine, isopropylamine, and ethylamine. The removal efficiency of 77.50% for chloride and 99.82% for sulfate was reported at the optimum conditions (pH: 8, temperature: 23.4°C, contact time: 26 min, and volume of solvent: 1.02). The recovered solvent also had the capacity for future application. Golbaz, Jafari, Rafiee, and Kalantary (2014) showed how to use coagulation to separate and remove phenol, Cr^{6+}, and CN^- from an aqueous solution. It was shown that the traditional coagulation method enhanced with ferric chloride successfully eliminated Cr^{6+}. The elimination of CN^- and phenol, however, was minimal. Additionally, the removal effectiveness for Cr^{6+} was significantly increased (up to 97%) as well as slightly enhanced for CN^- and phenol by raising the ferric chloride dose beyond 0.7 g/L. Increased SSA, a higher aggregation rate, and the breakdown of cyano complexes could all be contributing factors to the coagulation process' improved effectiveness.

4.2.3 BIOLOGICAL TREATMENTS

Recent developments in biochemistry, biotechnology, genetics, and microbiology have made biological treatment a competitive and effective technique for recovering water resources. The concentration of inorganics is reduced, COD and BOD organic loadings are reduced, and biological treatment is less expensive and environmentally beneficial. Colloidal and dissolved carbonaceous organic matter is converted by microbial treatment into settleable solids using substances like sulfate, nitrate, and carbon dioxide as electron acceptors. Biological treatments can be broadly divided into two categories: attached growth process and suspended growth process.

4.2.3.1 Suspended Growth Process

In the most widely utilized suspended growth method, activated sludge treatment, the soluble and insoluble organic and inorganic components are transformed into flocs by a dense microbial population in suspension. After that, gravity settling effectively eliminates the flocs of microorganisms and oxidized materials. To maintain the microbial concentration in the reactor, which in turn maintains the degree and rate of degradation, a portion of the sludge is recycled.

Activated sludge methods are used in one-step and multistep configurations, depending on how many steps are necessary for the sludge to decompose (Kim, Park, Lee, Jung, & Park, 2009). Plug flow activated sludge reactors, total mix activated sludge reactors, and sequencing batch reactors are a few examples of different design changes for activated sludge reactors (SBR). Different pH values, hydraulic residence times, and oxygen levels (absence or presence) are frequently needed for various contaminants, which increases the number of main reactors needed for degradation. Alternatives to high-maintenance treatment systems include waste stabilization ponds and artificial wetlands. They are capable of reducing composite mixes of organic chemicals that could bio-accumulate. Irrespective of the mode of degradation of the microbes, Monod's equation is used to evaluate the rate of substrate consumption:

$$\mu = \mu_{max} \frac{S}{K_S + S} \tag{4.1}$$

where
  μ = specific growth rate of the microorganisms,
  μ_{max} = maximum specific growth rate of the microorganisms,
  S = concentration of the limiting substrate for growth, and
  K_s = half-velocity constant

Surface-free water constructed wetlands, horizontal subsurface flow constructed wetlands, vertical subsurface flow constructed wetlands, and hybrid constructed wetlands are the several types of constructed wetlands utilized for industrial water resource recovery. By simultaneously removing heavy loadings of iron and manganese from discharged effluents, they have decreased freshwater use in steel factories (Xu et al., 2009). Built-in wetlands that rely only on algal populations have been utilized to remove discharged diseases and nutrients in addition to bacteria and fungi. The algae also prevent ammonia wastes from leaching by causing carbon dioxide fixation and incorporating them into the algal biomass. The research of the ammoniacal nitrogen (NH_3-N) removal process dynamics in an SBR using the Activated Sludge Model No. 1 (ASM1) and conventional SBR design computation for the best aeration time while matching the treatment criteria is reported in this paper. The study goes on to assess the effectiveness of NH_3-N removal using information from an existing SBR system. The reduction in aeration time from the existing 1.5 hours to 1.35 hours for 80% to 93% of NH_3-N removal resulted in a total energy savings of up to 10% for the computation of the SBR standard design. However, rehabilitation of the artificial wetlands and stabilization ponds takes a long time. Because of this, they are inadequate for continuously treating large amounts of wastewater. The substantial quantity of land required for scaling up these systems can be challenging in reproducing their performance at the laboratory scale.

A series of procedures, including SBR, up-flow anaerobic sludge blanket (UASB) reactor, and extended granular sludge beds, has been created as a result of advancements in suspended growth techniques. UASB reactors are employed in a variety of water resource recovery systems in the industry because they eliminate problems with anaerobic digesters, such as variable organic removal efficiency, and constant coliform count. By doing away with mechanical mixers, this procedure also saves on energy. Rising oxygen bubbles assure direct contact with the reactor's distinctive flocculant sludge, increasing turbulence and mass transfer rates. This enables independent variation of hydraulic retention times (Hickey, Wu, Veiga, & Jones, 1991). The primary degrading organisms in these systems are methanogenic bacteria, whose rate of degradation is accelerated by rising temperatures. It is frequently a multistage procedure with the addition of anaerobic filters and anaerobic hybrid reactors to improve the total removal efficiency (Chernicharo & Machado, 1998). The significant likelihood of microbial washout during shorter hydraulic retention times, however, is the fundamental disadvantage of operating a UASB reactor. It also has some drawbacks associated with microbiological systems, such as a lengthy startup phase, significant capital requirements, and so on.

4.2.3.2 Attached Growth Process

In such a procedure, the main reactor's media, which eliminates organics, nevertheless has bacteria adhering to it. The basic working principle of the system is the development of primarily aerobic bacteria on a medium like gravel, sand, or plastic screens, which act on the organics present in the wastewater entering the system. Trickling filters are one of the most popular attached growth techniques that have been used for recovering water resources. Higher hydraulic loadings and hazardous material can be managed since trickling filters provide a supportive environment for biofilm formation where microorganisms are retained for longer periods of time. Leca fraction columns, a type of dried expanded clay, were used in experiments to demonstrate a 100% nitrification rate and evaluate the efficacy of nitrifying filters (Kruner & Rosenthal, 2000). The three-phase fixed media reactors, characterized by the presence of a granular media filter and enable secondary treatment in the same unit, are a design improvement for attaining improved efficiency in the trickling filters (Mann & Stephenson, 1997). However, these reactors cannot handle the treatment of effluents with dissolved solids concentrations or changing shock loadings.

The effectiveness of separation is greatly increased when related biological treatment procedures are integrated with mechanical systems like fluidized bed reactors. The fluidized system gives all the benefits of a fluidized system over a packed bed reactor, including effective mixing and the capacity to operate continuously for a longer period of time. Packed bed reactors are also employed as an attached growth system. The microbial population is stabilized on glass beads and gravels in aerobic fluidized systems, which are then fluidized by an incoming jet of air or oxygen. Zeolite has also been considered a potential support component for microbial aerobic fluidized bed reactors.

Pozo-Antonio (2014) investigated how well nitrifying bacteria removed ammoniacal nitrogen from the environment. The microbial layer that forms inside breaks down the nutrients in the wastewater when it enters through the bottom. High separation efficiency, homogeneous wastewater distribution, and consistent mixing are all made possible by the fluidization process. After that, treated effluent is removed through a top outlet while the microbiological layer is occasionally rinsed off. Studies have revealed that aerobic fluidized bed reactors are effective in removing cyanide, color (89%), COD (83.3%), and sulfide (Petrozzi & Dunn, 1994). Low flow rates are associated with better separation efficiencies, but the amount of biomass attached to the fluidization medium directly relates to separation efficiency. An aerobic fluidized bed reactor has two modes of operation depending on whether the goal is to maximize the removal of pollutants or minimize the volume of sludge produced at the conclusion of the operation.

4.3 ADVANCED TREATMENT TECHNIQUES

4.3.1 ELECTROCOAGULATION

Wastewater from both home and industrial sources can be effectively treated electrochemically through a process called electrocoagulation (EC). The applied voltage difference across the electrodes in the EC process destabilizes the pollutants' charge and causes anode breakdown, releasing metal ions that act as coagulants and generating floc formation, which is then removed by precipitation or flotation (Tahreen, Jami, & Ali, 2020). Metal scraps were utilized as electrodes in a study by Vignesh, Siddarth, and Babu (2017) to remediate simulated electroplating effluent. The effects of changing operational parameters, including pH, current density (CD), initial Ni ion concentrations (500, 750, and 1000 ppm), EC time, and subsequent treatment efficiency responses, were noted. The removal efficiencies for Ni ions with Fe and Al electrodes were 95.9% and 94.1%, respectively, at the CD 30 mA/cm^2 and the EC time of 60 min. The metal scrap anode's reaction kinetics were enhanced by its high active surface area and porosity, which led to higher efficiency. In terms of cost and environmental sustainability, the metal scrap anode had the potential to perform better than the traditional anodes. Al-Shannag, Al-Qodah, Bani-Melhem, Qtaishat, and Alkasrawi (2015) also looked at the removal of Cu^{2+}, Cr^{3+}, Ni^{2+}, and Zn^{2+} ions from wastewater used in metal plating. They discovered that during 45 min of EC time, at a CD 4 mA/cm^2 and pH 9.56, more than 97% of heavy metal ions were removed. For an applied CD of 1, 2, 3, and 4 mA/cm^2, the elimination efficiency ranges in the EC process were 23–29%, 52–62%, 75–83%, and 98–100%, respectively. However, the EC procedure revealed no discernible variation in the removal effectiveness above a particular electrical conductivity (8.9 mS/cm). Furthermore, the CD was limited to 4 mA/cm^2 with EC duration between 45 and 60 min in order to concurrently maximize power consumption and removal efficiency. In a different investigation, 100% elimination of heavy metals (Cu, Cr, and Ni) was accomplished at a pH of 2.42, CD of 50 A/m^2, and EC duration of 30 min (Beyazit, 2014). The longer

CD and EC times might be responsible for the improved elimination of contaminants seen in the trials. This is because there is a direct correlation between the direct current field and potential electrolysis, which means that more metallic ions are released, leading to the production of more metal hydroxides and coagulants (Beyazit, 2014; Vignesh et al., 2017).

Malakootian and Heidari (2018) conducted a study to assess the viability of increasing organic matter removal by phenol elimination using a combined EC and PF procedure. The results of the experimental experiments carried out at a CD of 1.5 mA/cm^2 and pH of 4 revealed that the system's COD and phenol removal efficiencies were 100% and 99%, respectively. Contrary to other investigations, a decline in removal effectiveness was seen as the CD increased, which was consistent with the electrodes' polarization and inactivity. Furthermore, it was discovered that an optimal CD value is needed for pollutant removal after evaluating energy and electrode usage in the EC process.

4.3.2 MEMBRANE-BASED TECHNOLOGIES

Membrane technology has emerged in recent years as a desirable multi-step, sophisticated industrial water resource recovery solution. Membranes have a number of advantages, including improved separation efficiency, modular system design, operational flexibility, and simple process conditions to maintain (Purkait, Mondal, & Chang, 2019; Purkait, Singh, Mondal, & Haldar, 2020). The replacement of the extensive equipment utilized in the treatment of steel industry wastewater is a viable solution. The wastewater discharge regulations for integrated iron and steel plants are shown in Table 4.1. Steel effluents are appropriate for treatment by reverse osmosis and nano-filtration, which are pressure-driven membrane processes that separate charged solutes due to the presence of ions and their high conductivity (Mondal & Purkait, 2017; Mondal, Samanta, Meghnani, & Purkait, 2019).

Reverse osmosis and nano-filtration are two processes that have been used to treat effluents and have demonstrated 97% separation efficiency, the removal of heavy metal ions, BOD, COD, and total solids, as well as oil and grease, while also being environmentally benign. Comparing membrane procedures to more traditional separation methods like evaporation and distillation, the latter uses less energy (Purkait, Sinha, Mondal, & Singh, 2018a, b, c).

This method has been used in a variety of industries, including the recovery of enzymes from a fine chemical process, upgrading the gas stream in a bulk chemical process, and recovering iso-propanol (Meindersma & Kuczynski, 1996). According to a survey by Zheng et al. (2015), 6.7 million m^3 of wastewater are treated daily by around 580 membrane treatment process units in various industries, including the steel, petrochemical, and power generation sectors. The high organic-loaded effluents from blast furnaces, coke ovens, and SMS, as well as the high inorganic-loaded wastes from steel rolling mills, were subjected to independent fouling experiments by Choi, Yi, Moon, Sung, and Kang (2015). Their research showed that ultrafiltration tanks had severe biofouling that required

TABLE 4.1

Integrated Iron and Steel Plant: Wastewater Discharge Standards (Data from Das et al., 2018, Copyright © John Wiley and Sons)

Plant	Parameter	Concentration (mg/L, except pH)
Coke Oven by-Product Plant	pH	6.0–8.0
	Suspended Solids	100
	Phenol	1
	Cyanide	0.2
	BOD, 3 days, 27°C	530
	COD	250
	Ammoniacal Nitrogen	50
	Oil and Grease	10
Sinter Plant, Rolling Mills, and Steel Melting Shop	pH	6.0–9.0
	Suspended Solids	100
	Oil and Grease	10
Blast Furnace	pH	6.0–8.5
	Suspended Solids	50
	Oil and Grease	10
	Cyanide	0.2
	Ammoniacal Nitrogen	50

pre-disinfection. Including a pre-filtration phase like ultrafiltration or microfiltration, the concentration polarization issue that arises in membrane treatment processes can be considerably mitigated. Reverse osmosis and back washable microfiltration together have been shown to remove conductivity by around 95%, which is a direct estimate of solids removal (Lee, Kwon, & Moon, 2006).

A comprehensive treatment is provided by membrane filtration in conjunction with a method for stabilizing and discarding the concentrates. According to Huang, Ling, Xu, Feng, and Li (2011), a built wetland and an integrated ultrafiltration-reverse osmosis process resulted in a recovery rate of 75% and desalination up to 98%. With the inclusion of the created wetland, the hollow fiber poly-vinylidene fluoride (PVDF) polyamide composite membrane required less frequent backwashing. Pore blocking is one of the crucial elements in a membrane system that affects the process' separation flux. By gradually closing off the useful area provided for filtration, solid material in the feed increases filtration resistance. As a result, the filtrate flow rate decreases, necessitating a rise in total driving force. Evaluation of the reduced pore size, maximum filtrate volume, and fouling potential uses governing equations for constant pressure and constant flux separations in the interstices of a membrane (Iritani, 2013).

Amosa, Jami, Alkhatib, and Majozi (2016) used three different ultrafiltration membranes with various molecular weight cut-offs to conduct fouling investigations on a Newtonian fluid (bio-treated POME) (MWCOs). With four steps of total blocking, standard blocking, intermediate blocking, and cake filtration, which correspond to blocking indices ranging from 2 to 0, they used the blocking index, a dimensionless filtration constant that describes the type of fouling model, to measure the extent of fouling. The oil and grease components found in effluents can be handled by membrane techniques as well. Ultrafiltration membranes with pore diameters corresponding to 50,000 to 200,000 MWCO were demonstrated by Cheryan and Rajagopalan (1998) to be capable of creating permeates with an oil concentration as low as 10 ppm. The recovery of surfactants from the oil and grease content of steel sludges via membrane methods, such as microfiltration, has the potential to improve process economics.

According to Amosa et al. (2016), a cost-effective integrated PAC ultrafiltration bench scale method was technically feasible for facilitating water recovery in the last stage of wastewater treatment in the palm oil industry. They created the adsorbent for upstream treatment using a discarded fruit bunch and then managed the process downstream using a cross-flow polyethersulfone ultrafiltration membrane system made up of membranes with three different MWCOs of 1, 5, and 10 kDa. In combination with an ultrafiltration membrane with a 1 kDa MWCO, powdered activated carbon (PAC) was able to create permeate water that met U.S. EPA standards for boiler-feed and cooling water reuse. Smol, Wodarczyk-Makua, Bohdziewicz, and Mielczarek (2014) removed harmful PAHs from biologically treated coke oven wastewater using integrated coagulation, nano-filtration, and reverse osmosis membrane techniques.

According to comparisons of the removal efficiency of PAHs for the individual processes in the aforementioned integrated scheme, coagulation, nanofiltration, and reverse osmosis, respectively, had removal efficiencies of 38%, 68%, and 90%. Additionally, water recovery is made possible via membrane separation, which supports water reuse and recycling. New and potential technological advancements in this area include integrated systems that combine membrane separation processes with others, like membrane adsorption and membrane crystallization.

4.3.3 Advanced Oxidation Processes

4.3.3.1 Electrochemical Processes

These are a group of unit operations that include electricity as one of their constituent parts. Hazardous biorefractory compound levels are decreased through electrochemical oxidation, which also lowers BOD and COD loadings. Aluminum, iron, or hybrid Al/Fe electrodes are frequently used electrode materials. The electrochemical technique is frequently combined with established procedures such as coagulation (electro-coagulation), flocculation (electro-flocculation), dialysis (electro-dialysis), and COD removal. The efficacy of electro-coagulation and hybrid associates declines gradually in the following order: peroxi electro-coagulation > peroxi photo-electrocoagulation > photo-electrocoagulation > electrocoagulation. Electrochemical processes frequently result in the production of hydroxide ions, which expands the surface area available for the adsorption of organic ions and colloidal particles from the substrate. Consequently, electro-floatation makes it simple to separate insoluble flocs. Studies utilizing simulated wastewater showed that treatment effectiveness is directly correlated with the amount of time spent treating the wastewater and that the size of the electrochemical system is inversely correlated with the amount of energy passed. The problems caused by coke oven effluent, in particular, and also effluent from the steel industry, have been successfully solved by electrochemical oxidation. Graphite, lead dioxide-coated titanium (PbO_2/Ti), binary Ru-Ti oxide coated titanium, and tertiary Sn-Pd-Ru oxide coated titanium were the four electrodes examined by Chiang, Chang, and Wen (1995). Due to its resistance to phenolic oligomer adsorption, the PbO_2/Ti electrode had the highest COD removal effectiveness, at 89.5%. Additionally, it showed that the ammonium component, another unpleasant component of coke oven wastewater, was completely removed.

Another approach that has recently gained popularity and been industrially scaled up is electro-sorption. Electric voltage, time, pH, temperature, and electrode capacity are the factors that control how well the electro-sorption process works. Because of its high surface area for the adsorption of ions, activated carbon fiber is a frequently utilized electrode material. This material is particularly helpful when the heavy metal level of wastewater is around 200 ppm (Huang & He, 2013). In China, WISCO (Wuhan Iron and Steel Group Corporation) conducted an experiment employing electro-sorption that showed conductivity was reduced by 70% and calcium and chloride ions were removed by 75% and 68%, respectively, over the course of continuous operation at a high flow rate of 1000 L/h. Additionally, the primary danger posed by electrodes is the unintentional creation of toxic gas mixtures such as CCl_4, AsH_3, NO_x, SbH_3, and H_2S as well as dangerous mixes like H_2/O_2 and H_2/Cl_2. Therefore, rigorous design optimization is essential for these systems to operate safely. Furthermore, the method is energy-intensive and frequently economically undesirable due to the need for constant electricity.

4.3.3.2 Photo Degradation and Fenton Process

Due to its many benefits, including powerful radical-based reactions, high removal rate, total destruction of contaminants, recycling potential, and visible light harvesting, photo-degradation has been widely employed in wastewater treatment (Majumder, Saidulu, Gupta, & Ghosal, 2021). The catalyst is activated by UV radiation (photons) from a source, which leads to the creation of electron (e^-) and hole (h^+) pairs. While e^- reacts with the oxygen on the surface to form •O^{2-}, the generated h^+ reacts with H_2O to form •OH radicals. These radicals and holes break down the contaminants, resulting in the production of H_2O, CO_2, and other altered by-products (Gupta, Gupta, Ghosal, & Tiwary, 2020). Biswas et al. (2020) looked into the photocatalytic degradation of CN^- from steel industry effluent using H_2O_2 and UV irradiation. The rate of light deterioration peaked during the first 30 min then gradually decreased after that. After 3 hours, there was less than 1 mg/L of CN^- in the effluent, which means that more than 95% of the CN^- had been removed. The study found that 8 L/m^3 and 640 W, respectively, were the best H_2O_2 dosing rate and UV lamp power. The individual application of H_2O_2 and UV irradiation proved inefficient in eliminating CN^- from wastewater, in contrast to the combined application. A study examined the effectiveness of UV treatment to treat coking wastewater with an initial PAH concentration of 323 g/L. A 50% decrease in the concentration of PAHs was seen after the test samples were exposed to radiation for 30, 60, and 90 sec. Additionally, a total of 15 PAHs were found in the sample, with naphthalene having the greatest concentration. Additionally, the PAHs' individual removal efficiency ranged from 5% to 67%. Organics in wastewater can be effectively broken down by the Fenton process (Hansson, Kaczala, Marques, & Hogland, 2015; Silva & Baltrusaitis, 2021). Teixeira, Vieira, Yokoyama, and da Fonseca (2015) investigated the phenol removal efficiency to treat simulated coal processing wastewater with a starting phenol concentration of 200 mg/L and a pH range of 5–9. Steel wool, which contains ZVI, was used in the study as a potential catalyst for oxidizing phenol by H_2O_2 in wastewater. Tests showed that the phenol concentration in the wastewater was less than 0.5 mg/L at the optimal setting and a reaction time of 120 min, providing a removal efficiency of 99.75%. Additionally, it was shown that the reaction time could be shortened from 120 to 60 min by raising the temperature from 25°C to 45°C. The additional increase in H_2O_2 concentration and the Fe^0 surface area to volume ratio did not improve the phenol removal efficiency after satisfying the stoichiometric parameters for the initial stage of reaction kinetics. Additionally, iron-rich waste steel slag has been successfully used as a possible catalyst in the Fenton process to break down dyes and lower COD, supporting the repurposing and recycling of sludge (Ali, Gad-Allah, & Badawy, 2013; Heidari, Soleimani, & Mirghaffari, 2018).

4.3.3.3 Ozonation

To reach the appropriate amount of treatment, ozonation is a pH-dependent treatment method in which the targeted pollutants are reacted with either directly by molecular ozone (direct oxidation) or indirectly by free radicals (indirect oxidation) (Cunha, da Silva, Coutinho, & Marques, 2022; Joseph et al., 2021). Das, Anweshan, and Purkait (2021a) investigated the efficacy of standalone ozonation and electrocoagulation processes for the treatment of Cold Rolling Mill (CRM) effluent from the Tata Steel Industry in India, which had phenol, COD, BOD, iron, and oil content. To examine its impact on the removal efficiency of each process individually, the operational factors of the two standalone processes, such as current density (electrocoagulation), ozone formation rate (ozonation), and treatment time, were addressed. The amount of pollutant content could be reduced to below the corresponding discharge limits with the help of the ideal experimental settings of 200 A m^{-2} (current density), 1.12 mg s^{-1} (ozone generation rate), and 30 min (treatment time). According to the study, solitary electrocoagulation was shown to be more effective than ozonation in terms of cost and the effectiveness of pollutant removal for the intended effluent. Additionally, it was discovered that the ozonation procedure was nearly six times more expensive than electrocoagulation.

4.3.4 HYBRID TECHNOLOGIES

In order to tackle the complex mixture of toxic and hazardous substances present in steel industry wastewater, hybrid technologies have emerged due to their excellent treatment efficiency over other standalone techniques for treating steel plant wastewater. Nanofiltration (NF) reject effluent obtained from TATA Steel Industries, India, was utilized by Changmai et al. (2020) to study the removal efficiency of various elements such as magnesium, sodium, manganese, potassium, calcium, sulfate, and iron. In this study, electrocoagulation (EC) followed by a microfiltration (MF) process was adopted for the treatment study (Figure 4.2). Linz-Donawitz (LD) slag, a by-product from Tata Steel, was utilized for preparing ceramic membranes.

The membrane was then utilized for separating the flocks generated after electrocoagulation. It was observed that variations of parameters such as current density, operating time, and electrode distance played a vital role in the treatment process. The optimized conditions for current density, electrode distance, and operating time were found to be 50 Am^{-2}, 0.005 m, and 20 min, respectively. At optimized conditions, the concentration of Mg and Ca were found to drastically decrease to 54 and 18 mg L^{-1}, respectively, whereas Mn and Fe were completely removed from the effluent. In order to confirm the removal of metals through the integrated process, the electro-coagulated flocs were analyzed using EDX. The treated water was then recycled back to the other unit operations of the steel industry.

Similarly, an integrated ozonation-assisted electro-coagulation process was adopted by Das et al. (2021b) to treat the biological oxidation treatment unit effluent of the steel plant. The highest removal efficiencies obtained from the hybrid process for CN$^-$, COD, and BOD were 99.8%, 94.7%, and 95%, respectively. The removal efficiencies of the hybrid process, when combined with the standalone EC process, were found to produce better results. The study also revealed that current density and ozone generation rate were the predominant parameters which controlled the removal rate of CN$^-$. The removal efficiency got enhanced by 17.3%

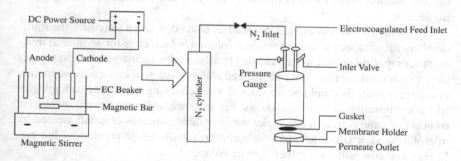

FIGURE 4.2 Schematic diagram of the electrocoagulation process followed by microfiltration for removal of flocks generated (Reproduced with permission from Changmai et al. 2020, Copyright © Taylor & Francis).

when the ozone generation rate increased from 1.0 to 1.33 mg/s. Moreover, it was also reported that after 40 min of EC operation, the removal efficiency of CN^- became constant, and no further removal occurred. Hence from the study, it was concluded that the hybrid integrated process of ozonation followed by EC was more efficient and economical than the standalone ozonation process.

In a related work by Kabdas li et al. (2010), organic matter and heavy metals were removed from complexed metal plating effluent using the integrated EC and Fenton method. According to the test results, the combined method removed heavy metals and organics with removal efficiencies of 100% and 70%, respectively. Additionally, it was shown that using the combination procedure increased organic matter removal by 20%. According to the studies mentioned above, CD is the main element controlling the EC process. Therefore, CD optimization is necessary for effective and cost-effective wastewater treatment. By combining EC with tertiary methods like the membrane process, it is possible to remove contaminants below the allowable discharge limits without producing potentially dangerous byproducts. The construction of membranes using slag from the steel industry will minimize overall waste generation and encourage recycling, and using solar energy as an energy source can further improve the system's sustainability.

4.4 CONCLUSION AND FUTURE PERSPECTIVES

The choice of an effluent treatment method should be based on the optimization of a number of factors, including toxicity levels, discharge standards, pollutants (their concentration and treated volume), and effluent volume. This is crucial for steel plants since the alloy steel preparation units produce effluents with various compositions and heavy metals that must meet strict discharge regulations. After administering tailored pre-treatments to ensure that treated effluents bear identical quality parameters, research must be done to produce a broad-spectrum technology for treating effluents from various sections. As a result of the current system's complexity and cost, which results in treated effluent with radically diverse quality, plans for overall water recycling are extremely challenging to implement.

In treatments like ion exchange, care has also been taken to reduce the introduction of ionic species into the residual solution, which calls for additional treatment. Comparing membrane technology to all complex industrial wastewater, it is clear that it is on the rise. The final treated water usage goal and a socioeconomic cost-benefit analysis should be considered when choosing membrane technologies like microfiltration or reverse osmosis. For instance, effluents treated by reverse osmosis or nano-filtration can make water potable, whereas effluents treated by microfiltration or Ultrafiltration would conform to discharge regulations but not. Energy and cost issues must therefore be balanced.

Solar cells and other alternative energy sources can be combined to power the separation process, resulting in even more energy savings and environmental advantages. However, the stabilization of membrane concentrates must be done before disposal, and membrane treatment is a physical separation phase that must be supplemented with chemical or biological treatments. By making investments in waste treatment profitable through product recovery and regeneration of raw materials using newer applications like membrane crystallization, it will be possible to choose processes that produce water with greater purity. Conventional low-cost treatment approaches, such as coagulation, can be combined with membrane-based separation techniques to provide an integrated system that is reliable, easy to use, and generates water of the highest quality that can be reused with the least amount of sludge possible.

REFERENCES

Ali, M. E. M., Gad-Allah, T. A., & Badawy, M. I. (2013). Heterogeneous Fenton process using steel industry wastes for methyl orange degradation. *Applied Water Science*, 3, 263–270.

Al-Shannag, M., Al-Qodah, Z., Bani-Melhem, K., Qtaishat, M. R., & Alkasrawi, M. (2015). Heavy metal ions removal from metal plating wastewater using electrocoagulation: Kinetic study and process performance. *Chemical Engineering Journal*, 260, 749–756.

Amosa, M. K. (2015). Process optimization of Mn and H₂S removals from POME using an enhanced empty fruit bunch (EFB)-based adsorbent produced by pyrolysis. *Environmental Nanotechnology, Monitoring and Management*, 4, 93–105.

Amosa, M. K., Jami, M. S., Alkhatib, M. F. R., & Majozi, T. (2016). Technical feasibility study of a low-cost hybrid PAC-UF system for wastewater reclamation and reuse: A focus on feed water production for low- pressure boilers. *Environmental Science and Pollution Research International*, 23, 22554–22567.

Amosa, M. K., Jami, M. S., Jami, M. S., Alkhatib, M. F. R., Jimat, D. N., & Muyibi, S. A. (2014). Comparative and optimization studies of adsorptive strengths of activated carbons produced from steam- and CO₂-activation for BPOME treatment. *Advances in Environmental Biology*, 8, 603–612.

Amosa, M. K., Jami, M. S., Muyibi, S. A., Alkhatib, M. F., & Jimat, R. D. N. (2013). Zero liquid discharge and water conservation through water reclamation & reuse of biotreated palm oil mill effluent: A review. *International Journal of Academic Research*. https://doi.org/10.7813/2075-4124.2013/5-4/A.24.

Amuda, O., Amoo, I., Ipinmoroti, K., & Ajayi, O. (2006). Coagulation/Flocculation process in the removal of trace metals present in industrial wastewater. *Journal of Applied Sciences and Environmental Management*, 10, 1–4.

Beh, C. L., Chuah, T. G., Nourouzi, M. N., & Choong, T. (2012). Removal of heavy metals from steel making waste water by using electric arc furnace slag. *E-Journal of Chemistry*, 9, 2557–2564.

Beyazit, N. (2014). Copper(II), chromium(VI) and nickel(II) removal from metal plating effluent by electrocoagulation. *International Journal of Electrochemical Science*, 9, 4315–4330.

Biswas, P., Bhunia, P., Saha, P., Sarkar, S., Chandel, H., & De, S. (2020). In situ photodecyanation of steel industry wastewater in a pilot scale. *Environmental Science and Pollution Research*, 27, 33226–33233.

Changmai, M., Das, P. P., Mondal, P., Pasawan, M., Sinha, A., Biswas, P., Sarkar, S., & Purkait, M. K. (2020). Hybrid electrocoagulation–microfiltration technique for treatment of nanofiltration rejected steel industry effluent. *International Journal of Environmental Analytical Chemistry*, 102(1), 62–83.

Chernicharo, C. A. L., & Machado, R. M. G. (1998). Feasibility of the UASB/AF system for domestic sewage treatment in developing countries. *Water Science and Technology, 38,* 325–332.

Cheryan, M., & Rajagopalan, N. (1998). Membrane processing of oily streams. wastewater treatment and waste reduction. *Journal of Membrane Science, 151,* 13–28.

Chiang, L. C., Chang, J. E., & Wen, T. C. (1995). Indirect oxidation effect in electrochemical oxidation treatment of landfill leachate. *Water Research, 29,* 671–678.

Choi, K.-J., Kima, S.-G., & Kimb, S.-H. (2008). Removal of antibiotics by coagulation and granular activated carbon filtration. *Journal of Hazardous Materials, 151,* 38–43.

Choi, S. K., Yi, H., Moon, J., Sung, Y., & Kang, S. G. (2015). Fouling characteristics of UF and RO membranes for reclamation of the wastewater from iron and steel industry. *International Journal of Environmental Science and Technology, 5,* 709–716.

Cunha, D. L., da Silva, A. S. A., Coutinho, R., & Marques, M. (2022). Optimization of ozonation process to remove psychoactive drugs from two municipal wastewater treatment plants. *Water, Air, & Soil Pollution, 233,* 67.

Das, P., Mondal, G. C., Singh, S., Singh, A.K., Prasad, B., Singh, K. K. (2018). Effluent Treatment Technologies in the Iron and Steel Industry - A State of the Art Review. *Water Environment Research.* 90 (5) 395–408.

Das, P. P., Anweshan, A., & Purkait, M. K. (2021a). Treatment of cold rolling mill (CRM) effluent of steel industry. *Separation and Purification Technology, 274,* 119083.

Das, P. P., Anweshan, Mondal, P., Sinha, A., Biswas, P., Sarkar, S., & Purkait, M. K. (2021b). Integrated ozonation assisted electrocoagulation process for the removal of cyanide from steel industry wastewater. *Chemosphere, 263,* 128370.

Gao, L., Shulei, L., Wang, Y., Gui, X., & Xu, H. (2016). Pre-treatment of coking wastewater by an adsorption process using fine coking coal. *Physicochemical Problems of Mineral Processing, 52,* 422–436.

Golbaz, S., Jafari, A. J., Rafiee, M., & Kalantary, R. R. (2014). Separate and simultaneous removal of phenol, chromium, and cyanide from aqueous solution by coagulation/precipitation: Mechanisms and theory. *Chemical Engineering Journal, 253,* 251–257.

Gupta, B., Gupta, A. K., Ghosal, P. S., & Tiwary, C. S. (2020). Photo-induced degradation of bio-toxic ciprofloxacin using the porous 3D hybrid architecture of an atomically thin sulfur-doped g-C3N4/ZnO nanosheet. *Environmental Research, 183,* 109154.

Hansson, H., Kaczala, F., Marques, M., & Hogland, W. (2015). Photo-Fenton and Fenton oxidation of recalcitrant wastewater from the wooden floor industry. *Water Environment Research, 87,* 491–497.

Heidari, B., Soleimani, M., & Mirghaffari, N. (2018). The use of steel slags in the heterogeneous Fenton process for decreasing the chemical oxygen demand of oil refinery wastewater. *Water Science and Technology, 78,* 1159–1167.

Hickey, R. F., Wu, W.-M., Veiga, M. C., & Jones, R. (1991). Start-up, operation, monitoring and control of high-rate anaerobic treatment systems. *Water Science and Technology, 24,* 207–255.

Huang, C.-C., & He, J.-C. (2013). Electro-sorptive removal of copper ions from wastewater by using ordered mesoporous carbon electrodes. *Chemical Engineering Journal, 221,* 469–475.

Huang, X. F., Ling, J., Xu, J. C., Feng, Y., & Li, G. M. (2011). Advanced treatment of wastewater from an iron and steel enterprise by a constructed wetland/ultrafiltration/reverse osmosis process. *Desalination, 269,* 41–49.

Iritani, E. (2013). A review on modeling of Pore-blocking behaviors of membranes during pressurized membrane filtration. *Drying Technology, 31,* 146–162.

Jami, M., Rosli, N., & Amosa, M. (2016). Optimization of manganese reduction in biotreated POME onto 3A molecular sieve and clinoptilolite zeolites. *Water Environment Research, 88,* 566–576.

Joseph, C. G., Farm, Y. Y., Taufiq-Yap, Y. H., Pang, C. K., Nga, J. L. H., & Li Puma, G. (2021). Ozonation treatment processes for the remediation of detergent wastewater: A comprehensive review. *The Journal of Environmental Chemical Engineering, 9,* 106099.

Kabbashi, A. N., Karim, M. I. A., Saeed, M. E., & Yacoob, K. H. K. (2008). Application of carbon nanotubes for removal of copper ion from synthetic water. *Biomed, 21,* 77–81.

Kim, Y. M., Park, D., Lee, D. S., Jung, K. A., & Park, J. M. (2009). Sudden failure of biological nitrogen and carbon removal in the full-scale pre-denitrification process treating cokes wastewater. *Bioresource Technology, 100,* 4340–4347.

Kruner, G., & Rosenthal, H. (2000). Efficiency of nitrification in trickling filters using different substrates. *Aquacultural Engineering, 2,* 49–67.

Lee, C. S., Robinson, J., & Chong, M. F. (2014). A review on application of flocculants in wastewater treatment. *Process Safety and Environmental Protection, 92,* 489–508.

Lee, J. W., Kwon, T. O., & Moon, I. S. (2006). Performance of polyamide reverse osmosis membranes for steel wastewater reuse. *Desalination, 189,* 309–322.

Li, H., Li, B. Q., Shao, H., & Zhang, D. (2011). Study on the treatment of steel wastewater with high turbidimetric by Boron Mud Haydite. *Applied Mechanics and Materials, 71–78,* 1282–1286.

Majumder, A., Saidulu, D., Gupta, A. K., & Ghosal, P. S. (2021). Predicting the trend and utility of different photocatalysts for degradation of pharmaceutically active compounds: A special emphasis on photocatalytic materials, modifications, and performance comparison. *Journal of Environmental Management, 293,* 112858.

Malakootian, M., & Heidari, M. R. (2018). Removal of phenol from steel wastewater by combined electrocoagulation with photo-Fenton. *Water Science and Technology, 78,* 1260–1267.

Mann, A. T., & Stephenson, T. (1997). Modelling biological aerated filters for wastewater treatment. *Water Research, 31,* 2443–2448.

Meindersma, G., & Kuczynski, M. (1996). Implementing membrane technology in the process industry: Problems and opportunities. *Journal of Membrane Science, 113,* 285–292.

Minhalma, M., & De Pinho, M. N. (2004). Integration of Nanofiltration/Steam stripping for the treatment of coke plant ammoniacal wastewaters. *Journal of Membrane Science, 242,* 87–95.

Mondal, P., & Purkait, M. K. (2017). Green synthesized iron nanoparticle-embedded pH-responsive PVDF-co-HFP membranes: Optimization study for NPs preparation and nitrobenzene reduction. *Separation Science and Technology, 52*(14), 2338–2355.

Mondal, P., Samanta, N. S., Meghnani, V., & Purkait, M. K. (2019). Selective glucose permeability in presence of various salts through tunable pore size of pH responsive PVDF-co-HFP membrane. *Purification Technology, 221,* 249–260.

Nowotny, N., Epp, B., von Sonntag, C., & Fahlenkamp, H. (2007). Quantification and modeling of the elimination behavior of ecologically problematic wastewater micropollutants by adsorption on powdered and granulated activated carbon. *Environmental Science & Technology, 41,* 2050–2055.

Petrozzi, S., & Dunn, I. J. (1994). Biological cyanide degradation in aerobic fluidized bed reactors: Treatment of almond seed wastewater. *Bioprocess Engineering, 11,* 29–38.

Pozo-Antonio, S. (2014). Determination of the zeolite optimal diameter for the settlement of nitrifying bacteria in an aerobic bed fluidized reactor to eliminate ammonia nitrogen. *Dyna, 81,* 21–29.

Purkait, M. K., Sinha, M. K., Mondal, P., & Singh, R. (2018a). *Interface Science and Technology,* Elsevier, Pages 115–144.

Purkait, M. K., Sinha, M. K., Mondal, P., & Singh, R. (2018b). Chapter 2 - pH-Responsive Membranes, Singh, *Interface Science and Technology*, Elsevier, Volume 25, Pages 39–66, ISBN 9780128139615.

Purkait, M. K., Sinha, M. K., Mondal, P., & Singh, R. (2018c). Chapter 4 - Photoresponsive Membranes, Editor(s): Mihir Kumar Purkait, Manish Kumar Sinha, Piyal Mondal, Randeep Singh, Interface Science and Technology, Elsevier, Volume 25, Pages 115–144, ISBN 9780128139615.

Purkait, M. K., Mondal, P., & Chang, C.-T. (2019). Treatment of Industrial Effluents: Case Studies, CRC Press, 1st Edition, ISBN 9781138393417.

Purkait, M. K., Singh, R., Mondal, P., & Haldar, D. (2020). Thermal Induced Membrane Separation Processes, Elsevier, ISBN 9780128188019.

Samanta, N. S., Banerjee, S., Mondal, P., Anweshan, Bora, U., & Purkait, M. K. (2021). Preparation and characterization of zeolite from waste Linz-Donawitz (LD) process slag of steel industry for removal of Fe^{3+} from drinking water. *Advanced Powder Technology*, 32(9), 3372–3387.

Servais, P., Billen, G., & Bouillot, P. (1994). Biological colonization of granular activated carbon filters in drinking-water treatment. *Journal of Environmental Engineering*, 120, 888–899.

Silva, M., & Baltrusaitis, J. (2021). Destruction of emerging organophosphate contaminants in wastewater using the heterogeneous iron-based photo-Fenton-like process. *The Journal of Hazardous Materials*, 2, 100012.

Smol, M., Wodarczyk-makua, M., Bohdziewicz, J., & Mielczarek, K. (2014). The use of integrated membrane systems in the removal of selected pollutants from pre-treated wastewater in coke plant. *Membranes and Membrane Processes in Environmental Protection Monographs of the Environmental Engineering Committee Polish Academy of Sciences*, 119, 143–152.

Tahreen, A., Jami, M. S., & Ali, F. (2020). Role of electrocoagulation in wastewater treatment: A developmental review. *Journal of Water Process Engineering*, 37, 101440.

Teixeira, L. A. C., Vieira, N. D. A., Yokoyama, L., & da Fonseca, F. V. (2015). Degradation of phenol in mine waters using hydrogen peroxide and commercial steel wool. *International Journal of Mineral Processing*, 138, 15–19.

Vazquez, I., Rodriguez-Iglesias, J., Maranon, E., Castrillon, L., & Alvarez, M. (2007). Removal of residual phenols from coke wastewater by adsorption. *Journal of Hazardous Materials*, 147, 395–400.

Vignesh, A., Siddarth, A. S., & Babu, B. R. (2017). Electro-dissolution of metal scrap anodes for nickel ion removal from metal finishing effluent. *Journal of Material Cycles and Waste Management*, 19, 155–162.

Xu, J. C., Chen, G., Huang, X. F., Li, G. M., Liu, J., Yang, N., & Gao, S. N. (2009). Iron and manganese removal by using manganese ore constructed wetlands in the reclamation of steel wastewater. *Journal of Hazardous Materials*, 169, 309–317.

Zheng, X., Zhang, Z., Yua, D., Chena, X., Chenga, R., Mina, S., Wang, J., Xiao, Q., & Wang, J. (2015). Overview of membrane technology applications for industrial wastewater treatment in China to increase water supply. *Resources, Conservation and Recycling*, 105, 1–10.

5 Treatment of Ironmaking Wastewater

5.1 OVERVIEW OF IRONMAKING WASTEWATER

Steel is produced in three ways around the world: the blast furnace-basic oxygen furnace route (BF-BOF), electric arc furnace route (EAF), and direct reduced iron-electric arc furnace route (DRI-EAF). The Blast furnace route primarily consumes three raw materials: iron ore, coal, and coke. A sinter strand and a coke oven are used to process iron ore and coal, respectively. The hot metal formed in the blast furnace is then sent to the basic oxygen furnace (BOF), where the carbon content is reduced, and molten steel is formed (Changmai et al., 2022; Perpiñán et al., 2023).

Blast Furnace is a countercurrent reactor that uses oxygen-enriched hot air and auxiliary fuel that is injected through tuyeres in the lower section of the furnace together with a solid burden, ferrous materials, and coke that are charged from the top of the furnace. The solid burden descends inside the BF while the reducing gases ascend. The iron oxides are dried, heated, and then reduced by the ascending reducing gases as the ferrous materials descend in the furnace. The chemical processes that take place in the BF are extremely complex (Lu, Pan, & Zhu, 2015).

Many gases are produced during the iron-making process and are later expelled from the top of the blast furnace. Typically, the gas passes through air and water cleaning systems as it passes through gas cleaning equipment. After the particle matter is removed, the gas is cooled, and the purified gas is used as energy to produce iron. The gas transfers numerous different types of solid particles from the blast furnace to the gas cleaning system. The blast furnace's load materials and operating conditions affect how much gas the existing components in the furnace emit, and this has an impact on the wash water's quality. A dust bag and a water cleaning system remove the gas' coarser and finer constituents, respectively. Solids in the wash water settle to the bottom of a settling tank, where it is conveyed. The solid components are processed as sludge in centrifuges, and the purified water is then recirculated to the gas cleaning step (Kiventerä, Leiviskä, Keski-Ruismäki, & Tanskanen, 2016).

Water spraying is used to clean up blast furnace gas and remove particles. Coke is quenched with seawater. Carbon monoxide, carbon dioxide, soot, nitrogen oxides, and chlorides make up the majority of the BF flue gases, which go through a gas-cleaning process. There are two stages in this process: primary and secondary. The first stage includes a dust catcher where all heavy metals are removed. In the secondary step of a wet cleaning procedure, all suspended particles in the gas are removed and dissolved in water while the gas is being cleaned

DOI: 10.1201/9781003366263-5

in contact with water. Clean BF gas is used to heat furnaces after the secondary
stage, and the polluted water from the system contains a lot of dissolved salts,
particularly chlorides and sulfates.

Ore particles, dust, cyanides, sulfur and phenol compounds, slag, ash content,
and metal ions along with other dissolved solids are found in the wastewater.
Because of its porous structure, seawater used for coke quenching can be identified as an origin for diffusing chloride into coke (Mukherjee, Mondal, Sinha,
Sarkar, & De, 2016; Purkait, Sinha, Mondal, & Singh, 2018a, b, c. This chloride
diffused coke is used as a raw material in the ironmaking process, increasing the
chloride content of the blast furnace flue gas discharge.

5.2   TREATMENT TECHNIQUES FOR THE TREATMENT
OF BLAST FURNACE EFFLUENT

The flow and characteristics of the blowdown vary greatly between the steel
industry due to various factors. Recycling water in the gas washing system possibly leads to scaling within the system. Different research opposing such viewpoints have been studied, and various approaches to scaling control have been
proposed, including CO_2 retention in the recycle, acid addition, and also adding
of scaling inhibitors such as organic phosphonate compounds. A typical blowdown treatment includes coagulation with various coagulants, such as ferric salts
and organic polymer, followed by breakpoint chlorination. Further, a combined
biological treatment with coke-plant effluents, after some physicochemical pre-
treatment, has also been proposed.

One of the steel industry in India, for example, employs chemical coagulation and biological methods. This method is very effective for the removal of
cyanide and phenol contents from the effluent. After the Biological method, the
treated water still contains high levels of ions such as chlorides, sulfates, and
fluorides, making it difficult to reuse or discharge into the water streams (Deepti
et al., 2020). It is reported that the main source of such ions in the effluent is
the use of seawater for coke quenching. As a result, the aforementioned industry
employs nanofiltration (NF) as a treatment technique, with a particular emphasis
on chloride removal. Nanofiltration is preferable to other techniques including
Reverse Osmosis because it can operate at lower transmembrane pressures as well
as works effectively in the removal of monovalent and divalent ions (Chakraborty,
Purkait, DasGupta, De, & Basu, 2003; Purkait, Kumar, & Maity, 2009; Yaranal,
Kumari, Narayanasamy, & Subbiah, 2019).

Nanofiltration could reduce the ion concentration in the effluent. The rejected
stream of this process, on the other hand, is concentrated with chlorides, sulfates,
sodium, calcium, magnesium, and potassium ions. Furthermore, the rejected
stream can account for up to 30% of the feed. This rejected stream cannot be
reused or discharged into the environment because it may interfere with the activities of the environment. Chlorides (> 600 mg/L) and sulfates (> 400 mg/L),
in particular, cause adverse effects on living beings, such as diarrhea, dehydration, and hypertension (Usepa, 2002). Scaling, pipe blockage, corrosion, and

demineralization, on the other hand, are some of the industry's negative impacts. Thus, it is imperative to treat the nanofiltration rejected stream before reusing or subleasing it to the environment.

5.3 TREATMENT TECHNIQUES FOR HIGHLY SALINE WASTEWATER

The most popular means of managing nanofiltration/reverse osmosis rejected streams are deep well injection, evaporation ponds, sewer discharge, and oceanic outfall (Semblante, Lee, Lee, Ong, & Ng, 2018). These approaches, however, are no longer practicable due to environmental restrictions and stringent discharge standards. As a result, several researchers attempted to develop a process for treating highly saline-rejected brine stream of NF/RO processes. Electrodialysis, integrated system, vacuum membrane distillation, forward osmosis (Mondal & Purkait, 2017; Mondal, Samanta, Meghnani, & Purkait, 2019; Purkait, Mondal, & Chang, 2019; Purkait, Singh, Mondal, & Haldar, 2020), ion exchange, geothermal and solar desalination, and biosorption are among the treatment techniques considered (Bader, 1998; Dolar, Košutić, & Strmecky, 2016; Mericq, Laborie, & Cabassud, 2010). However, these technologies are not economically viable due to the excessive energy and space requirements. Furthermore, the limited output and poor removal efficiency make it unsuitable for industrial use.

5.3.1 SOLVENT-BASED PRECIPITATION

Precipitation is an old separation process technique that is gaining popularity in various applications due to its high efficiency. Miscible organic solvents can precipitate salts due to their unique phenomenon called solventing out. Solventing out is nothing but the reduction in salt solubility (Bader, 1994). Miscible organic solvents are advantageous because they i) can cause high salt precipitation, ii) have a boiling point close to ambient temperature, iii) are less expensive, and iv) pose less environmental risk. Furthermore, solvents are widely used as herbicides in agriculture (Mansour, 1995).

In comparison to conventional methods, such as distillation, the fast vaporization of organic solvents using a vacuum is becoming more and more feasible and economical. Isopropylamine, diisopropylamine, and ethylamine are among these organic solvents having the aforementioned characteristics. These solvents also have additional qualities such as 1) superior species blending, 2) improved dose and transport control, 3) improved species separation, and 4) adaptability to varied situations (Deepti et al., 2021).

5.4 CASE STUDY

The practical treatment setup for the treatment of highly saline brine from the nanofiltration rejected stream from the steel industry's blast furnace process is detailed in this section. Miscible organic solvents such as diisopropylamine,

isopropylamine, and ethylamine are used in various ratios in the precipitation process as a treatment technique. It deals with the experimental setup of the process along with the process parameters optimization. A real-life effluent from nanofiltration was taken for the experiment, and the results were discussed in order to give a better insight into the experimental variations.

5.4.1 EXPERIMENTAL SECTION

This section mainly deals with the materials used for the experimental runs along with the methodology utilized for carrying out experiments. It discusses several techniques for analyzing various process variables and reasonable explanations for the outcomes of the study.

5.4.1.1 Materials Used for the Study

The effluent for the study (Nanofiltration rejected water) was collected from Tata Steel Ltd, India. The effluent was tightly sealed and kept in a dark chamber at room temperature to avoid any change in the composition caused by oxidation. Hydrochloric acid, sodium hydroxide, chloride salts of sodium, potassium, calcium, iron, magnesium, and manganese were purchased from Merck (India). Miscible organic solvents, such as Isopropylamine (IPA), diisopropylamine (DIPA), and ethylamine (EA), were procured from Spectrochem Pvt.Ltd in Mumbai, India.

5.4.1.2 Analytical Methods

Ion chromatography was used to analyze the chloride and sulfate ions in the sample (Metrohm Ltd, Herisau, 792 basic IC). Iron, magnesium, manganese, sodium, potassium, and calcium ions were analyzed using an atomic absorption spectrophotometer. Borosil, India provided all the glassware utilized in the experiments. Using a microprocessor water/soil analysis kit, the pH, conductivity, and TDS were measured (Model: VSI-301, Make: VSI Electronics). Using Fourier transform infrared spectroscopy (Model No.: IRAffinity-1; Make: M/s Shimadzu, Japan), functional group identification was analysed. The crystallinity of precipitated salts was evaluated by X-ray diffraction (Model No. D8 Advance, Make: Bruker, the Netherlands). Using field emission scanning electron microscopy, the morphological characteristics of the precipitated salt were examined (FESEM, make: Zeises, model: Sigma). The elements of the precipitated salt were determined using energy-dispersive X-ray spectroscopy.

5.4.1.3 Experimental Method Used for the Treatment Process

In a ceramic jar using an agitator, the effluent sample and solvent in various proportions were thoroughly mixed. The solvent-to-sample ratio (VR), which may be determined using Eqn. 5.1, was taken into consideration. The solution was mixed for 1 hour at an ambient temperature at 100 RPM. After 30 min, blending was stopped for an additional 30 min to allow the salt to settle. Further, to remove the salts from the solution, vacuum filtration was used to filter the solution through nylon filter paper (47 mm diameter; pore size of 0.2 μm). A 50-Hz vacuum pump with a 15 L/min flow rate was used. The organic solvent was recovered by passing the filtrate (solvent-water mixture) to a distillation unit with a capacity of 1 L. For later usage, the distillate (organic solvent) was then stored. Figure 5.1 depicts the experimental design in detail. Using Eqn. 2, it was determined how effectively the two ions were separated in terms of the percentage precipitation factor.

$$VR = \frac{\text{Volume of solvent}}{\text{volume of sample}} \qquad (5.1)$$

$$\%P = 1 - \left(\frac{Cf}{Ci}\right) \times 100 \qquad (5.2)$$

where VR is the volume ratio, %P, Ci, and Cf is the percentage precipitation factor, initial anion concentration in the initial sample (effluent), and final anion concentration in the filtered sample (after treatment), respectively.

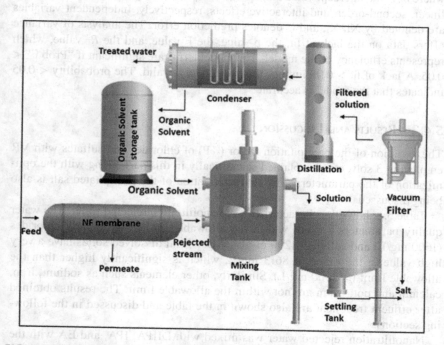

FIGURE 5.1 Detailed layout of precipitation process (Reproduced with permission from Deepti et al., 2020, Copyright © Elsevier).

5.4.1.4 Experimental Design and Process Parameter Optimization

Response surface methodology (RSM) is a set of statistical and mathematical methods for evaluating the impact of various factors and their actions on system response. The primary benefit of RSM is the ability to develop and optimize variables, resulting in fewer experimental runs. RSM is primarily concerned with experimental design, response surface modelling, and finally, optimization (Mondal & Purkait, 2018; Navamani Kartic, Aditya Narayana, & Arivazhagan, 2018). The central composite design (CCD) was used, with four independent factors, namely volume ratio, pH, temperature, and contact time, and two dependent responses, namely percentage precipitation factor (% P) of chloride and sulfate. The independent variables are represented by X1, X2, X3, and X4. The range of variables is as follows: VR = 0.3–2.6, pH = 2–10, temperature = 10–32°C, contact time = 10–30 min. The interaction between the independent variable and the dependent responses is shown in Eqn. 5.3.

$$\gamma = b_0 + \sum_{i=1}^{k} b_1 X_1 + \sum_{I=1}^{K} b_{ii} X_i^2 + \sum_{i=1}^{k-1} \sum_{J=1}^{K} b_{ij} X_i X_j + C \tag{5.3}$$

where b_0 is the intercept value and b_i, b_{ii}, and b_{ij} are the regression coefficients for linear, second-order, and interactive effects, respectively. Independent variables are denoted by Xi, Xj, and C denotes prediction error. The analysis of variance offers data on the lack of fit, the p-value, the F value, and the R^2 value, which represents efficiency of the model. The model terms are significant if "Prob F" < 0.05. A lack of fit > 0.05 implies that the model is valid. The probability < 0.05 indicates that the model is accurate.

5.4.2 RESULTS AND DISCUSSION

The variation of the precipitation factor (%P) of chlorides and sulfates with VR employing 3 solvents is explained scientifically in this part, along with the optimization of the parameters. The characterization of the precipitated salt is also being discussed.

Table 5.1 shows the characteristics of nanofiltration rejected water. The water quality parameters are not within the allowable limits, particularly chlorides (1600 mg/L) and sulfates (4200 mg/L). Also, Total dissolved solids have a very high value in the range of 8613 mg/L, which is significantly higher than the allowable limit (> 2000 mg/L). Similarly, other elements such as sodium, iron, calcium and potassium are not within the allowable limit. The results obtained after effluent treatment are also shown in the table and discussed in the following sections.

Nanofiltration rejected water was mixed with DIIPA, IPA, and EA with the volumetric proportion of solvent to rejected water (VR). All the experiments were conducted at room temperature with 10 and 30 min blending and settling times, respectively. Figures 5.2a and 5.2b show the variation of precipitation factor with

TABLE 5.1

Characteristics of NF Rejected Water (Reproduced with Permission from Deepti et al., 2020, Copyright © Elsevier)

Parameters	NF rejected water	After treatment	Permissible limit for surface water
Chloride (mg/L)	1594	358	600
Sulfate (mg/L)	4196	8.4	400
Total Dissolved Solids (mg/L)	8613	785	2000
pH	8.1	7.25	9.0
Turbidity (NTU)	0.5	0.3	140
Sodium (mg/L)	650	142	200
Potassium (mg/L)	175	4	12
Calcium (mg/L)	183	51	200
Iron (mg/L)	38	4.2	3
Magnesium (mg/L)	198	122	150
Manganese (mg/L)	21	14	2

VR for chloride and sulfate, respectively, for the three solvents considered here. The precipitation factor for chloride increases sharply up to a VR value of 1.25 and then gradually thereafter, as shown in Figure 5.2a. The increase in precipitation factor is marginal above the VR value of 3, which can be seen from the figure. Again, precipitation is greater for DIIPA and less for IPA and EA for each value of VR. Given an example, for VR = 2.67, chloride precipitation is 80.46% for DIIPA, 69.54% for IPA, and 67.65% for EA (Figure 5.2a). A comparable observation is made for sulfate precipitation, as illustrated in Figure 5.2b. However, unlike chloride, the gradual increase is observed up to VR of 1.0 and then becomes steady.

The highest precipitation factor for chloride (80%) and sulfate (100%), as seen in Figures 5.2a and 5.2b, is attained by using DIIPA. The difference in the degree of salt precipitated was caused by the different hydrogen bonding capacities of the solvents. Salts are salted out when molecules of water are readily H-bonded with amine-based solvents and are not accessible for salt solvation. Diisopropylamine has a greater capability for H-bonding than the other two solvents. As a result, depending on H-bonding capacities, the degree of chloride and sulfate precipitation is in the order *DIIPA > IPA > EA* (Finar, 1963). In terms of operation and boiling point, IPA is the optimum choice (Perry & Green, 1984). As a result, isopropylamine was preferred for examining numerous precipitation-influencing factors, which will be detailed in the next sections.

Models for the precipitation factors for chloride and sulfate are provided by Eqns. 5.4 and 5.5, respectively. Studies were conducted using a volume ratio, VR (A), a range of pH (B), a range of temperature (C), and a range of contact duration (D). A positive coefficient value denotes a synergistic impact, while a negative coefficient value represents the reverse effect (Deepti et al., 2020).

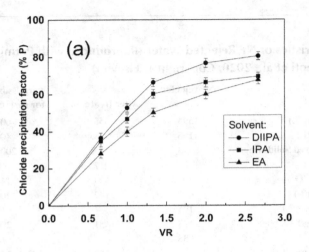

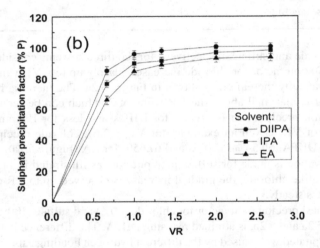

FIGURE 5.2 Variation of precipitation factor (%P) with VR (a) For chloride precipitation, and (b) For sulfate precipitation [Reproduced with permission from Deepti et al. 2020, Copyright © Elsevier].

Chloride Precipitation factor $(\%P) = 76.68 - 3.04 \times A - 0.91 \times B - 3.82$
$\times C - 1.69 \times D + 1.44 \times A \times B + 4.48 \times A \times C + 0.79 \times A \times D + 0.75$ $\qquad(5.4)$
$\times B \times C - 1.86 \times B \times D - 4.05 \times C \times D - 2.25 \times A^2 - 5.03 B^2 \times C^2 - 0.050 \times D^2$

Sulfate Precipitation $(\%P) = 98.44 + 0.73 \times A + 0.040 \times B - 0.064 \times C - 0.50$
$\times D - 0.18 \times A \times B - 0.46 \times A \times C + 0.65 \times A \times D - 0.12 \times B \times C - 0.35$ $\qquad(5.5)$
$\times B \times D + 0.46 \times C \times D + 6.32 \times A^2 - 0.052 \times B^2 + 0.052 \times C^2 + 0.17 \times D^2$

In the above equations, the coded values are the representations for the variables: A= volume ratio (VR), B = pH, C = temperature (°C), and D = time (min). The impacts of the primary four independent parameters have been studied using a total of thirty experiments with six central points. The responses were analyzed, with the findings of an ANOVA for the precipitation of chlorides and sulfates. With an F value of 172.8, the low probability indicates the reliability of the model. The model is also very viable as the lack of fit (0.08) is more than 0.05. From ANOVA outcomes, it was observed that $R^2 = 0.99$ and an adjusted $R^2 = 0.98$ indicate that the model is valid for chloride precipitation. Likewise, an F value of 18 or less with a low probability suggests that the model is reliable. The model's lack of fit, which is 0.82 and greater than 0.05, shows that it is reliable. The sulfate precipitation model has a significant R^2 value of 0.98.

The influence of the dependent variables on the percent precipitation factor for chlorides and sulfates are studied. The study helps to connect the relation between (i) VR and pH, temperature, contact time, (ii) pH and temperature, contact time, and (iii) temperature - contact time. Additionally, this shows the precipitation factors at their maximum (77.50% for chloride and 99.82% for sulfate) and minimum (49.48% for chloride and 94.21% for sulfate) as a result of the aforementioned interactions. A rise in VR and pH resulted in an increase in the chloride precipitation factor. The intermediate value of the design space had the highest chloride precipitation (63.05%). The result made it abundantly evident that, as VR and pH increase, the contact period has little to no effect on the chloride precipitation factor. However, the temperature and dosage significantly influenced the chloride precipitation factor. From the results obtained, it was concluded that the best mixing time for precipitating salts in the presence of isopropylamine at an increasing volume ratio (VR) is 15 min. Similar to this, it was also seen that when VR increases, the sulfate precipitation factor increases up to 99%. The impact of temperature and VR on the sulfate precipitation factor demonstrates that at higher VR (2.08) and lower temperature (15.50°C), the sulfate precipitation factor reaches its maximum (98.5%). Also, it was envisaged from the results that increasing temperature and pH had no effect on the sulfate precipitation factor. However, the percentage of chloride and sulfate precipitation factors are better around the midpoints of the limit. The results of the interaction research provide a clear picture of how the four elements under consideration affect variations in the precipitation component. A study on the interactions between pH and VR was conducted using two pH values, 4 and 8, changing VR and maintaining contact times and temperatures at 20 and 21°C, respectively. Two things can be gathered from the nitration study. First, chloride precipitation is higher (80%) at lower VR while the sample's pH is kept at 4. The chloride precipitation factor, however, eventually dropped to 72% as VR was raised. On the other hand, at pH 8, the chloride precipitation factor was 74.5%, which remained constant until VR 1.50 before declining. Second, whereas the chloride precipitation factor drastically decreases at pH 4, it does not significantly reduce at pH 8 with varied VR. This would suggest that maintaining the sample pH at its initial value (8.1) improves the elimination of chloride ions. The chloride precipitation factor gradually decreases from 82.2% to 65% at 15°C.

At 26.5°C, on the other hand, the VR-dependent chloride precipitation factor increased up to 68%. Since the temperature is close to that of room temperature and precipitation is more likely at this temperature, there is no need for additional energy, which directly benefits in terms of cost-effectiveness. Similarly, the chloride precipitation factor exhibits a decreasing trend relative to an increasing VR when the contact period is between 14.0 and 25.0 min.

The pattern in the sulfate precipitation factor interaction study is exactly the reverse of the chloride precipitation factor. As the VR increases for pH 4 and 8, respectively, the sulfate precipitation factor rises from 97.15% to 99.25% and from 97.9% to 98.5%. Also, the sulfate precipitation factor rises progressively with rising VR at 15°C and 26°C. However, when the temperature is held at 15°C, the sulfate precipitation factor is maximum (99.55%). This suggests that sulfate ions are less soluble than chloride ions at lower temperatures.

The sulfate precipitation factor remained nearly constant at 99.25% with an increase in VR at 14.0 min of contact time. At 25.0 min, VR is increased from 0.92 to 2.09, and the sulfate precipitation factor jumps drastically up to 99.7%. Thus, it can be said that contact time longer than 14.0 min results in better sulfate precipitation factors. The interactive effects between the factors effecting the chloride and sulfate precipitation factor is also shown in Figures 5.3 and 5.4, respectively.

Numerical optimization was used to select a suitable value for each input and each response. The range of the inputs and the target of the responses were taken into consideration while setting an output value for the specified conditions (Dastkhoon, Ghaedi, Asfaram, Ahmadi Azqhandi, & Purkait, 2017; Mourabet, El Rhilassi, El Boujaady, Bennani-Ziatni, & Taitai, 2017). All input objectives were chosen for an option, "in range" and "maximum" for the obtained responses. At VR of 1.02, pH of 8, temperature of 23.39°C, and contact duration of 26 min, the highest percentages of chloride (77.50%) and sulfate (99.82%) precipitation factors were recorded. By doing the experiments at the ideal conditions recommended by the software, validation was verified. The confirmatory trials showed precipitation factors for sulfate as 97.6% and chloride as 74.50%, which indicates the accuracy of the model.

When recovery is taken into account, the solvents utilized for the process should have advantageous physical characteristics. Isopropylamine has a very low boiling point, which makes it easier to recover the solvent from the solution. With simple distillation, 96–98% of the solvent was recovered. However, evaporation throughout the procedure resulted in a 2–4% loss. Modern distillation plants might be able to prevent this loss. Traces of solvent from the filtered material were removed by the process of aeration.

5.4.3 CHARACTERIZATION OF PRECIPITATED SALT

Surface morphological aspects of the resultant precipitates were examined using field emission scanning electron microscopy (FESEM). It was seen from the FESEM images that the salts were agglomerated in nature. This is because a

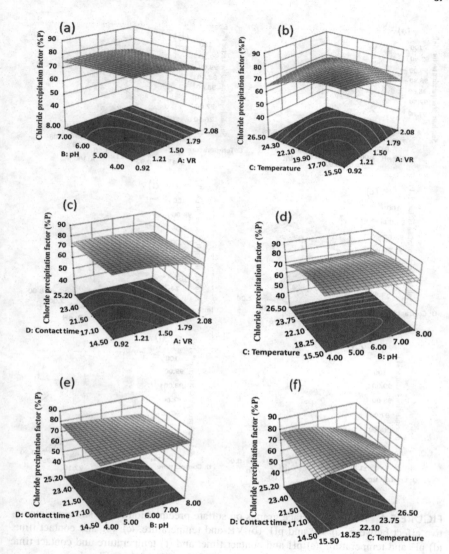

FIGURE 5.3 Response surface graphs on chloride precipitation factor showing interactive effect between (a) VR and pH, (b) VR and temperature, (c) VR and contact time, (d) pH and temperature, (e) pH and contact time, and (f) temperature and contact time [Reproduced with permission from Deepti et al., 2020, Copyright © Elsevier].

variety of inorganic salts are present within precipitated salts. Using EDX, a quantitative analysis of the salt sample was carried out by determining the elemental makeup of each sample species. From the analysis, it was seen that the precipitated salt comprises cations such as magnesium, calcium, sodium, iron, and potassium, which may react with chloride and sulfates to produce the corresponding salts.

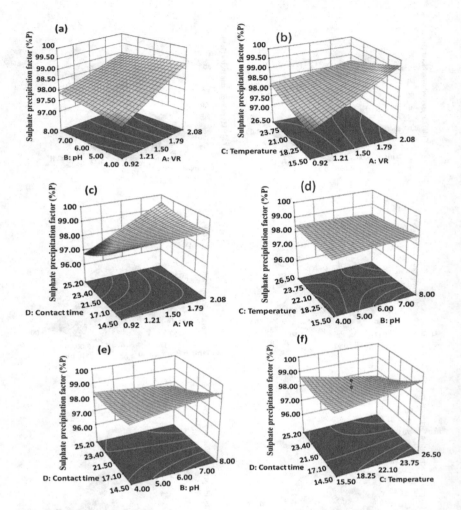

FIGURE 5.4 Response surface graphs on sulfate precipitation factor showing interactive effect between (a) VR and pH, (b) VR and temperature, (c) VR and contact time, (d) pH and temperature, (e) pH and contact time, and (f) temperature and contact time [Reproduced with permission from Deepti et al., 2020, Copyright © Elsevier].

The results of X-Ray diffraction (XRD) analysis of precipitated salt are explained below. The peaks corresponding 31° and 45° indicated the presence of sodium chloride (Boopathy & Sekaran, 2017). Similar to this, 30° and 43.4°, respectively, indicated the presence of calcium sulfate and magnesium chloride (Favvas, Stefanopoulos, Vordos, Drosos, & Mitropoulos, 2016; Zhang, Lu, Pan, Wang, & Yang, 2013). Thus, the outcomes show that the available cations in the solution have salted out the chloride and sulfate. The precipitated salt was also analyzed using Fourier transform infrared spectroscopy (FTIR) at wavelengths between 500 and 4000 cm^{-1}. Inorganic sulfates such as calcium sulfate, sodium sulfate, and potassium sulfate were present in precipitated salt, with concentrations ranging from 1000 to 1110 cm^{-1}. In addition, N-H stretching was visible at peaks between 3694 cm^{-1} and 3377 cm^{-1}. This might be because the precipitated salt contains a very small quantity of IPA. It should be noted that because the lattice vibrations of inorganic chlorides are outside the IR range, they cannot vibrate; hence, inorganic chlorides cannot be seen in FTIR analysis (Derrick, M.R., Stulik, D. and Landry, 1999) (Derrick, Stulik, & Landry, 1999).

5.5 MODELING AND SCALE UP LIMITATIONS

To increase water recycling and reuse, early process feasibility studies seek to discover the most convenient solution. Process modeling and simulations are useful tools for identifying different solutions to increase water efficiency and analyzing available process integration (PI) options for water systems, taking into account equipment and plant capacities, efficiency, and quality for process streams. Since it can be challenging to conduct experimental tests in actual plants, scenario analyses can be developed through the use of modeling and simulations in virtual plants. Before they are tested and implemented at the plant level, solutions for enhancing resource management are evaluated through simulation. It can be viewed as a transversal tool that enables thorough evaluations taking into account all the important factors necessary to evaluate the viability of changing overly complex processes. Simulation can also be used to evaluate unexpected scenarios or operating circumstances that are difficult to assess and test. In terms of water management, simulations try to evaluate ways to reduce freshwater input prior to actual tests or applications by measuring the effects on the quality of the blowdown water. Despite using an optimization tool in the preceding study, a unit model for the proposed modular treatment process can be developed using mass and energy balances based on available experimental data or first-principles models from commercial software. In addition, an economic model can be combined with the process model to estimate capital and operational costs using economic assumptions. Optimization strategies would be developed based on the evaluation criteria to converge toward a design that maximizes water reuse, minimizes chemical and energy footprints of treatment, and increases process profitability by generating saleable by-products. The precipitation process is comprised of simple and easy-to-use operational equipment with a short running period. However, more research is required to develop a thermodynamic framework based on liquid-liquid equilibrium criteria in order to predict precipitation data. Deep study and investigations in light of volume ratio are required to lower the process's running costs. The majority of research studies have been conducted at the laboratory scale using artificially made solutions; therefore, pilot plant scale using real industrial effluent has to be encouraged to investigate the opportunity of using solvent-based precipitation for treatment of real industrial effluent.

5.6 SUMMARY

Chlorides and sulfates from nanofiltration rejected wastewater of the steel industry were successfully separated by miscible organic solvents such as diisopropylamine, isopropylamine, and ethylamine based on the solventing out phenomenon. Diisopropylamine was found to have a higher capacity for precipitating ions. From the results obtained, it can be easily noted that there was a huge reduction of Total Dissolved Solids, chlorides, and sulfates along with other ions, which are also within the permissible limits. Response Surface Methodology was very helpful in determining the optimized conditions for the precipitation of both chloride and sulfate ions. The paramount conditions such as volume ratio, temperature, contact time, and pH for chloride and sulfate precipitation were attained. Almost 98% of the recovered organic solvent (IPA) was reused in the system. However, introducing a closed system for the entire precipitation mechanism will increase recovery efficiency. Based on the findings of this investigation, it is strongly advised to conduct a pilot plant study to treat the nanofiltration rejected stream and to look into the viability of doing so on an industrial scale. The steel industry is urged to look deeper into the aforementioned technology for usage on an industrial scale if it already has an integrated plant of NF units for the treatment of blast furnace wastewater.

REFERENCES

Bader, M. S. H. (1994). Separation of salts from aqueous saline solutions: Modeling and experimental. *Journal of Environmental Science and Health. Part A: Environmental Science and Engineering and Toxicology*, 29(3), 429–465. https://doi.org/10. 1080/10934529409376047

Bader, M. S. H. (1998). Precipitation and separation of chloride and sulfate ions from aqueous solutions: Basic experimental performance and modelling. *Environmental Progress*, 17(2), 126–135. https://doi.org/10.1002/ep.670170220

Boopathy, R., & Sekaran, G. (2017). Studies on process development for the separation of sodium chloride from residue after evaporation of reverse osmosis reject solution. *Separation and Purification Technology*, 183(April 2017), 127–135. https://doi. org/10.1016/j.seppur.2017.04.008

Chakraborty, S., Purkait, M. K., DasGupta, S., De, S., & Basu, J. K. (2003). Nanofiltration of textile plant effluent for color removal and reduction in COD. *Separation and Purification Technology*, 31(2), 141–151. https://doi.org/10.1016/ S1383-5866(02)00177-6

Changmai, M., Das, P. P., Mondal, P., Pasawan, M., Sinha, A., Biswas, P., Sarkar, S., & Purkait, M. K. (2022). Hybrid electrocoagulation–microfiltration technique for treatment of nanofiltration rejected steel industry effluent. *International Journal of Environmental Analytical Chemistry*, 102(1), 62–83.

Dastkhoon, M., Ghaedi, M., Asfaram, A., Ahmadi Azqhandi, M. H., & Purkait, M. K. (2017). Simultaneous removal of dyes onto nanowires adsorbent use of ultrasound assisted adsorption to clean waste water: Chemometrics for modeling and optimization, multicomponent adsorption and kinetic study. *Chemical Engineering Research and Design*, 124(April 2017), 222–237. https://doi.org/10.1016/j. cherd.2017.06.011

Deepti., Bora, U., & Purkait, M. K. (2021). Promising integrated technique for the treatment of highly saline nanofiltration rejected stream of steel industry. *Journal of Environmental Management, 300*(July), 113781. https://doi.org/10.1016/j.jenvman.2021.113781

Deepti., Sinha, A., Biswas, P., Sarkar, S., Bora, U., & Purkait, M. K. (2020). Separation of chloride and sulphate ions from nanofiltration rejected wastewater of steel industry. *Journal of Water Process Engineering, 33*(December 2019), 101108. https://doi.org/10.1016/j.jwpe.2019.101108

Derrick, M. R., Stulik, D., & Landry, J. (1999). Scientific Tools for Conservation: Infrared Spectroscopy in Conservation Science. Paul Getty Trust, Los Angeles. https://doi.org/10.1002/9781118162897.ch5

Dolar, D., Košutić, K., & Strmecky, T. (2016). Hybrid processes for treatment of landfill leachate: Coagulation/UF/NF-RO and adsorption/UF/NF-RO. *Separation and Purification Technology, 168*(May 2016), 39–46. https://doi.org/10.1016/j.seppur.2016.05.016

Favvas, E. P., Stefanopoulos, K. L., Vordos, N. C., Drosos, G. I., & Mitropoulos, A. C. (2016). Structural characterization of calcium sulfate bone graft substitute cements. *Materials Research, 19*(5), 1108–1113. https://doi.org/10.1590/1980-5373-MR-2015-0670

Finar, I. (1963). I.L Finar-Vol 1.pdf.

Kiventerä, J., Leiviskä, T., Keski-Ruismäki, K., & Tanskanen, J. (2016). Characteristics and settling behaviour of particles from blast furnace flue gas washing. *Journal of Environmental Management, 172*, 162–170. https://doi.org/10.1016/j.jenvman.2016.02.037

Lu, L., Pan, J., & Zhu, D. (2015). Quality Requirements of Iron Ore for Iron Production, Iron Ore: Mineralogy, Processing and Environmental Sustainability, Elsevier Inc, Pages 476–504. https://doi.org/10.1016/B978-1-78242-156-6.00016-2

Mansour, S. B. (1995). Precipitation and separation of salts, scale salts, and norm contaminant salts from saline waters and saline solutions. U S patent 5468394.

Mericq, J. P., Laborie, S., & Cabassud, C. (2010). Vacuum membrane distillation of seawater reverse osmosis brines. *Water Research, 44*(18), 5260–5273. https://doi.org/10.1016/j.watres.2010.06.052

Mondal, P., & Purkait, M. K. (2018). Green synthesized iron nanoparticles supported on pH responsive polymeric membrane for nitrobenzene reduction and fluoride rejection study: Optimization approach. *Journal of Cleaner Production, 170*(August 2017), 1111–1123. https://doi.org/10.1016/j.jclepro.2017.09.222

Mondal, P., & Purkait, M. K. (2017). Green synthesized iron nanoparticle-embedded pH-responsive PVDF-co-HFP membranes: Optimization study for NPs preparation and nitrobenzene reduction. *Separation Science and Technology, 52*(14), 2338–2355.

Mondal, P., Samanta, N. S., Meghnani, V., & Purkait, M. K. (2019). Selective glucose permeability in presence of various salts through tunable pore size of pH responsive PVDF-co-HFP membrane. *Separation and Purification Technology, 221*, 249–260.

Mourabet, M., El Rhilassi, A., El Boujaady, H., Bennani-Ziatni, M., & Taitai, A. (2017). Use of response surface methodology for optimization of fluoride adsorption in an aqueous solution by brushite. *Arabian Journal of Chemistry, 10*(December 2013), S3292–S3302. https://doi.org/10.1016/j.arabjc.2013.12.028

Mukherjee, R., Mondal, M., Sinha, A., Sarkar, S., & De, S. (2016). Application of nanofiltration membrane for treatment of chloride rich steel plant effluent. *Journal of Environmental Chemical Engineering, 4*(1), 1–9. https://doi.org/10.1016/j.jece.2015.10.038

Navamani Kartic, D., Aditya Narayana, B. C. H., & Arivazhagan, M. (2018). Removal of high concentration of sulfate from pigment industry effluent by chemical precipitation using barium chloride: RSM and ANN modeling approach. *Journal of Environmental Management, 206*(October 2017), 69–76. https://doi.org/10.1016/j.jenvman.2017.10.017

Perpiñán, J., Peña, B., Bailera, M., Eveloy, V., Kannan, P., Raj, A., & Romeo, L. M. (2023). Integration of carbon capture technologies in blast furnace based steel making: A comprehensive and systematic review. *Fuel, 336*, 127074. https://doi.org/10.1016/j.fuel.2022.127074

Perry, R., & Green, D. (1984). *Perry's Chemical Engineer's Handbook*, McGraw-Hill. https://doi.org/10.1036/0071511253

Purkait, M. K., Kumar, V. D., & Maity, D. (2009). Treatment of leather plant effluent using NF followed by RO and permeate flux prediction using artificial neural network. *Chemical Engineering Journal, 151*(1–3), 275–285. https://doi.org/10.1016/j.cej.2009.03.023

Purkait, M. K., Sinha, M. K., Mondal, P., & Singh, R. (2018a). Interface Science and Technology, Elsevier, Pages 115–144.

Purkait, M. K., Sinha, M. K., Mondal, P., & Singh, R. (2018b). Chapter 2 - pH-Responsive Membranes, Singh, Interface Science and Technology, Elsevier, Volume 25, Pages 39–66, ISBN 9780128139615.

Purkait, M. K., Sinha, M. K., Mondal, P., & Singh, R. (2018c). Chapter 4 - Photoresponsive Membranes, Editor(s): Mihir Kumar Purkait, Manish Kumar Sinha, Piyal Mondal, Randeep Singh, Interface Science and Technology, Elsevier, Volume 25, Pages 115–144, ISBN 9780128139615.

Purkait, M. K., Mondal, P., & Chang, C.-T. (2019). Treatment of Industrial Effluents: *Case Studies*, CRC Press, 1st Edition, ISBN: 9781138393417.

Purkait, M. K., Singh, R., Mondal, P., & Haldar, D. (2020). Thermal Induced Membrane Separation Processes, Elsevier, ISBN: 9780128188019.

Semblante, G. U., Lee, J. Z., Lee, L. Y., Ong, S. L., & Ng, H. Y. (2018). Brine pre-treatment technologies for zero liquid discharge systems. *Desalination, 441*(April), 96–111. https://doi.org/10.1016/j.desal.2018.04.006

Usepa. (2002). Development document for final effluent limitations guidelines and standards for the iron ans steel manufacturing point source category. *Response*, (April), 1062.

Yaranal, N. A., Kumari, S., Narayanasamy, S., & Subbiah, S. (2019). An analysis of the effects of pressure-assisted osmotic backwashing on the high recovery reverse osmosis system. *Journal of Water Supply: Research and Technology-Aqua*, 1–21. https://doi.org/10.2166/aqua.2019.089

Zhang, Z., Lu, X., Pan, F., Wang, Y., & Yang, S. (2013). Preparation of anhydrous magnesium chloride from magnesium chloride hexahydrate. *Metallurgical and Materials Transactions B: Process Metallurgy and Materials Processing Science, 44*(2), 354–358. https://doi.org/10.1007/s11663-012-9777-5

6 Treatment of Biological Oxidation Treated Wastewater
Case Study

6.1 INTRODUCTION

Commonly, wastewater produced by steel manufacturers is poisonous, dangerous, and has an unpleasant odor and lasting hue. The quantitative aspect is typically used to evaluate the wastewater's environmental impact. The water quality of wastewater discharged, however, contributes more significantly to environmental harm. Tata Steel, Jamshedpur, uses 25 m³ of water on average for each ton of steel produced and produces 5 m³ of wastewater per hour of operation (Mukherjee, Mondal, Sinha, Sarkar, & De, 2016; Purkait, Sinha, Mondal, & Singh, 2018a, b, c). Various unit operations of the steel-making process, viz. Linz-Donawitz (LD) converters, smelting of pig iron in a blast furnace, rolling mills, and a number of other support activities, produce effluent with various chemical compositions. Biological and chemical coagulation techniques are typically utilized to remediate wastewater produced by the steel industries (Das et al., 2021c). Tata Steel, Jamshedpur (India), treats the effluent from coke oven byproducts using the biological oxidation method.

Wastewater that has undergone biological oxidation treatment (BOT) is particularly concerning since it contains a number of harmful contaminants, including cyanide, phenol, color, and COD. These effluents frequently contain cyanide, especially those from blast furnace outputs. It is important to avoid releasing cyanide-contaminated effluents into natural water bodies (Changmai et al., 2022; Das et al., 2021a). Less than 0.2 mg/L of cyanide may be discharged into surface waters. The presence of cyanide has deleterious effects on human health that can be both immediate and chronic. It deactivates the cytochrome oxidase enzyme and prevents oxygen-sensitive organs from utilizing oxygen for aerobic metabolism. The two main types of cyanide in wastewater are HCN and CN⁻ ion (Kariim, Abdulkareem, Tijani, & Abubakre, 2020). Additionally, phenol and its components play a significant role in the production of resin as well as in petrochemicals, medicines, plastics, ceramics, refineries, steel plants, and coal transformation techniques. Excessive levels of phenol in surface and potable water raise serious concerns for the health of people, plants, microbes, and animals. While the allowable level of phenol is 1 mg/L for surface water, its

DOI: 10.1201/9781003366263-6

permitted level is 0.5 µg/L for potable water (Das, Sharma, & Purkait, 2022b). In addition to these significant pollutants, the existence of residual color, produced by the steel industry's coke oven plant, is the main cause for concern (Das, Anweshan, & Purkait, 2021b). The primary issue for the aquatic ecology can also be caused by the discharge of colored effluent. Lignin is produced in wastewater when hard carbons from biomass waste or hardwood biomass are utilized in the steelmaking process. Thus, the lignin's chromophore groups are mostly to blame for the effluent's brownish hue. Additionally, the altered photosynthetic activity caused by the decreased penetration of solar radiation into water disturbs the ecosystem of flora and fauna. The maximum amount of color allowed in surface water is 300 Hazen (Sharma & Philip, 2016). Additionally, the Chemical Oxygen Demand (COD) and Biological Oxygen Demand (BOD) water quality indicators give a measure to assess the impact of wastewater released into naturally receiving water bodies. A significant amount of oxidizable organic material is present in high COD concentrations, which lowers the dissolved oxygen level and harms aquatic life. For surface water, the acceptable limit for COD is 250 mg/L, and BOD is 30 mg/L (Das, Sharma, & Purkait, 2022b). As a result, too many of these refractory components in BOT effluent pose a hazard to the biological treatment processes, as they may primarily suppress microbial activity.

Literature on the treatment of industrial wastewater using various methods are widely available. The most commonly employed of them are adsorption (Samanta, Das, Mondal, Changmai, & Purkait, 2022; Samanta et al., 2021), nanotechnology (Das et al., 2021a; Das, Mondal, & Purkait 2022a; Purkait, Mondal, & Chang, 2019; Purkait, Singh, Mondal, & Haldar, 2020), oxidation (Saravanan et al., 2022), membrane separation (Changmai et al., 2022; Mondal & Purkait, 2017; Mondal, Samanta, Meghnani, & Purkait, 2019), and biological approaches (Sharma, Das, Sood, Chakraborty, & Purkait, 2022) among others. Although the aforementioned processes are widely used, a thorough examination of all these techniques reveals several limitations connected to each of the processes. Furthermore, because of its complexity, each procedure is recognized to have limitations when it comes to the effective treatment of actual industrial effluents. As a result, there is a need for more efficient and cost-effective hybrid approaches that use fewer chemicals, use less energy, and have a higher capacity to remove contaminants. Advanced oxidation processes (AOPs), viz. UV/TiO$_2$/H$_2$O$_2$, UV/Fe^{2+}/H$_2$O$_2$, Ozone/electrocoagulation, electrolysis/Fenton, UV/Ozone, and photoelectro-Fenton, among others, have drawn significant attention owing to their quick reaction time, high oxidative ability, straightforward equipment design, and low operating costs (Dong et al., 2022; Ma et al., 2021).

This chapter contains a thorough discussion and summary of the ozone-assisted electrocoagulation procedure for treating BOT effluent from the Tata steel mill. Both the ozonation and electrocoagulation processes' experimental designs have been demonstrated. In-depth information has also been provided on the removal of contaminants such as cyanide, phenol, color, and

COD content at various current densities, electrode distances, and operational times. A study on the operating costs and energy usage of the combined ozone-based electrocoagulation process has also been added toward the end of this chapter.

6.2 MATERIALS AND EXPERIMENTAL METHODS

6.2.1 MATERIALS

To carry out the experiments for the current investigation, Tata Steel Ltd. Jamshedpur (India) provided the biological oxidation-treated wastewater produced after the coke oven process. To change the pH of the solution, 98% pure H_2SO_4 (Merck) and 0.1 M NaOH were utilized. For cleaning the electrodes, diluted HCl/H_2SO_4 (0.01 M) was utilized in order to maintain their effectiveness. Electrocoagulation was carried out using aluminum electrodes. Using 120-grit sandpaper, the electrodes were washed after every operation to prolong their efficacy. The analytical reagent (AR) grade chemicals were purchased from Merck and utilized directly in the tests. A photometer was used to assess the performance of the integrated ozone-electrocoagulation approach (Das et al., 2022b).

6.2.2 EXPERIMENTAL METHODS

6.2.2.1 Ozonation Set-Up

For the ozonation procedure, an ozone generator and oxygen concentrator were also employed. Ozone is normally produced at a rate of 0–3 mg/s. To produce microbubbles, ozone was injected into the reactor using a sparger. One liter of effluent was used for each analysis, and the glass reactor could hold up to two liters. With the aid of an ozone destructor, the surplus O_3 was converted into O_2 (Patel, Majumder, Das, & Ghosh, 2019). The ozonation procedure was carried out at three different ozone-generating rates: 1.33, 1.11, and 1.00 mg/s. However, it was discovered through testing that ozone production rates above 1.33 mg/s greatly increased the overall energy consumption and the operational costs of the entire process. At an ozone production rate of 1.33 mg/s, all the parameters treated were efficiently decreased below their given permitted limits. Additionally, continuing the ozonation process after the initial 40 min of operation produces results that are comparable. The ideal operating parameters were therefore determined to be an ozone production rate of 1.33 mg/s and a treatment period of 40 min (Das et al., 2021a). The sample analysis was performed every 5 min, and a photometer was used to measure the change in contaminant concentration. The iodometric approach was used to calculate the amounts of both ozone production and consumption. A simplified illustration of the combined ozone-assisted electrocoagulation process is shown in Figure 6.1.

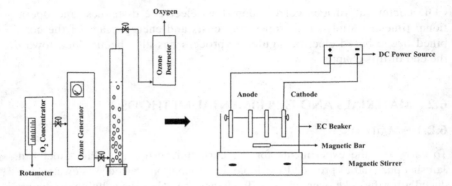

FIGURE 6.1   Schematic depiction of the integrated ozonation assisted electrocoagulation process (Reproduced with permission from Das et al., 2021c @ Elsevier).

6.2.2.2   Electrocoagulation Set-Up

The investigations were carried out in a 500 mL acrylic electrochemical reactor (batch system). Aluminum sheets of 7 cm × 3.5 cm in size and 24.5 cm² in effective surface area were employed as the electrode material for the cathode and the anode. The amount of wastewater used for each analysis was 300 mL. To deliver consistent current, a direct current (DC) power source was connected to the electrodes (total 4) via bi-polar configuration. Induced polarisation occurs when a potential is applied to the electrodes' ends, which causes the assembly as a whole to become bipolar (Ghosh, Medhi, & Purkait, 2008). The inter-electrode spacing remained constant at 0.005 m.Current densities of 150, 100, and 50 A/m² were selected for the electrocoagulation procedure. However, it was discovered that experiments with current densities greater than 100 A/m² produced results that were comparable to or barely reduced the quantity of pollutants when compared to experiments with 100 A/m². As a result, the experiment's ideal current density was determined to be 100 A/m². Similar to this, a 30-min electrocoagulation period was chosen as the ideal working period because continuing the trials past this point would only marginally reduce the amount of pollutants removed. A magnetic stirrer was utilized to maintain a uniform speed of 250 rpm to efficiently spread the coagulants created by metal dissolution from anodes. The flocs should not be broken by excessive stirring. According to the results of the research, swirling at speeds above 250 rpm results in a reduction in the ability of aluminum ions to form flocs, which lowers the process' removal effectiveness. Additionally, a reactor-based homogenous mixture could not be produced at stirring speeds lower than 250 rpm. Therefore, 250 rpm was taken as the optimum stirring speed. The temperature for all of the analyses was set to room temperature (Das et al., 2021a). The solution was then decanted for further examination after the electrocoagulated water was left overnight for the sludge to settle. Table 6.1 shows that using the electrocoagulation or the ozonization method alone was insufficient to remove the pollutants beneath their set allowable levels.

TABLE 6.1

Removal Efficacy of the Individual Ozonation and Electrocoagulation Process (Modified with Permission from Das et al., 2021a and Das et al., 2021c @ Elsevier)

Parameter	Initial content	Final content	Desirable level (WHO)
Ozonation Process (at optimal conditions)			
Cyanide (mg/L)	150	9	0.5
Phenol (mg/L)	150	11	1
Color (Hazen)	2450	250	300
COD (mg/L)	2050	570	250
Electrocoagulation process (at optimal conditions)			
Cyanide	150	57	0.2
Phenol	150	85	1
Color	2450	1950	300
COD	2050	815	250

6.3 RESULTS AND DISCUSSION

6.3.1 PERFORMANCE OF OZONE-ASSISTED ELECTROCOAGULATION PROCESS

6.3.1.1 Cyanide Removal

The efficiency of cyanide degradation, together with ozone production rate and ozonation time, was determined. It was found that the proportion of cyanide removed increased from 76.7% to 94.0%, respectively, as the rate of ozone formation increased from 1 to 1.33 mg/s. Furthermore, cyanide reduces from an initial content of 150 mg/L to 9 mg/L at an optimal ozone production rate and operational period of 1.33 mg/s and 40 min, respectively. When free cyanide ions come into contact with ozone, they quickly degrade. Ozone oxidation in the presence of cyanide undergoes the following reaction mechanism (Barriga-Ordonez, Nava-Alonso, & Uribe-Salas, 2006).

$$CN^- + O_3 \rightarrow CNO^- + O_2 \tag{6.1}$$

$$2CNO^- + 3O_3 + H_2O \rightarrow 2HCO_3^- + N_2 + O_2 \tag{6.2}$$

It is well known that the first reaction happens quickly while the slow and rate-determining phase is the cyanate oxidation via ozone. During the operation, ozone converts cyanide ions to cyanate while releasing O_2 gas. Cyanate ions are transformed into N_2 gas and carbonate ions through further oxidation. The cyanate ion oxidizes at a rate that is around one-fifth that of cyanide.

$$CNO^- + 2H_2O \rightarrow NH_3 + CO_2 + OH^- \tag{6.3}$$

The cyanate ion quickly undergoes acid hydrolysis at low pH to produce NH_3 and CO_2. The effluent pH was maintained in the acidic domain and reached a minimal value of 3 during the ozonation process, which encouraged the development of gaseous products. The exponential growth of the liquid phase is caused by the development of CO_2 and N_2 gases throughout the reaction (Rowley & Otto, 1980). Within the first 15 min of treatment, over three-fourths of the original cyanide concentration is oxidized (at an ideal ozone production rate of 1.33 mg/s). However, as the experiment continues, the rate of cyanide elimination slows. Eventually, after 40 min of operation, no appreciable decrease in cyanide concentration was seen. The rates of ozone formation have little bearing on this trend. Additionally, the percentage of cyanide elimination increases as the rate of ozone formation rises. Nonetheless, it was observed that ozonation alone could not bring the cyanide content beneath its allowable level of 0.2 mg/L.

The electrocoagulation procedure eliminated the residual cyanide using aluminum electrodes as cathode and anode. The process follows the sweep coagulation and/or surface adsorption mechanism. The cyanide concentration progressively drops during the electrocoagulation procedure from 9 mg/L (achieved at 1.33 mg/s) to 0.1 mg/L for an ideal electrolysis period and current density of 30 min and 100 A/m^2, respectively. Through the deposition process, the surface of aluminum hydroxyl flocs captures and eliminates colloidal and soluble pollutants from the aqueous solution. Trivalent aluminum ions (Al^{3+}) are released when aluminum electrodes are anodized, and a wide range of other polymeric and monomeric species, including $Al(OH)^{4-}$, $Al(OH)^{2+}$, $Al_8(OH)_{20}^{4+}$, $Al_2(OH)_2^{4+}$, $Al_{13}(OH)_{34}^{5+}$, and $Al_6(OH)_{15}^{3+}$ (Changmai et al., 2022). The generated hydroxyl ions and hydroxyl species are effective at eliminating the ionic pollutants. Therefore, the electrocoagulation procedure was able to significantly decrease the amount of cyanide ions present. The linear drop in cyanide content with electrocoagulation time is independent of changes in current density. Nonetheless, the maximum removal efficiency was produced by the use of high current density. Thus, for an electrolysis period of 30 min and a current density of 100 A/m^2, the highest degradation efficacy achieved throughout the process was 98.8%.

6.3.1.2 Phenol Removal

The ozonation of phenol produces the •OH radicals. The electron transfer from phenol/phenolate to ozone is thought to be the primary reason (Do & Chen, 1994).

$$PhO^-/PhOH + O_3 \rightarrow PhO^•/PhOH^{•+} + O_3^{•-} \qquad (6.4)$$

The production of an ozonide radical and a phenoxyl radical results from the direct transfer of an electron between ozone and phenol in Eqn. (4). While oxidizing organic chemicals into carbon dioxide and water, ozone oxidizes inorganic substances to their higher oxidation states. The rate of phenol oxidation will increase in direct proportion to the quantity of dissolved ozone in the wastewater because of an increased ozone production rate. According to the literature, oxidation degrades phenol during ozonation, producing catechol and hydroquinone at the para-position, which then quickly changes into o- and p-benzoquinones. According to several accounts, the oxidation of hydroquinone, which results in the synthesis of benzoquinone, is either caused by a direct attack via molecular ozone at the para-position, followed by loss of H_2O_2, or by a $HO_2^{\bullet}$ attack at the para-position of the phenoxyl radical, followed by H_2O loss (Ramseier & von Gunten, 2009). The pH drops during the ozonation process, which promotes the oxidation of benzoquinone to organic acids along with a decrease in benzoquinone yield. In other words, the aromatic rings open, which causes the breakdown of o- and p-benzoquinone into organic acids. H_2O and CO_2 are the byproducts of the oxidation of these organic acids. The phenol concentration is decreased as different phenolic byproducts start to develop and then oxidise into CO_2 and H_2O (Yang, Hu, Huang, & Yan, 2010). Thus, it may be inferred that as the ozonation experiment progresses, the amount of phenol contained in BOT effluent dramatically reduces. As a consequence, phenol falls from an initial content of 150 mg/L to a final content of 11 mg/L at an optimal ozone production rate of 1.33 mg/s and an operating time of 40 min. It was found that the percentage of phenol degradation likewise increases from about 79.5% to 92.6% as the ozone production rate enhances from 1 to 1.33 mg/s.

The phenol content further drops during the electrocoagulation process from 11 mg/L (achieved at 1.33 mg/s) to 0.5 mg/L at an ideal electrolysis period and current density of 30 min and 100 A/m², respectively. The elimination of phenol is observed to increase with the operating period and current density using Al electrodes. According to the literature, phenol, along with organic acids, reacts with Al^{3+} ions to form insoluble constituents via integrated coagulation, complexation, and/or precipitation phenomenon. In addition, phenol and other related organic molecules tend to go in the direction of an aqueous solution. It is believed that they are adsorbed onto the hydrolyzed products created throughout the process, which can then be confined inside the developing hydroxides (coagulation phenomenon), and the subsequent particle components can physically interact to produce flocs/sludge (flocculation phenomenon) (Dhadge, Medhi, Changmai, & Purkait, 2018). Additionally, phenol (along with other organic constituents) is reduced to very small molecules at the cathode with sufficient current density. These tiny organic molecules and the suspended particles are adsorbed by $Al(OH)_3$ flocs, which are created throughout the process and are then removed via H_2 flotation or precipitation (Ghosh, Medhi, & Purkait, 2011). As such, it can be determined that a rise in current density causes an increase in bubble density and a reduction in its size, intensifying the degradation efficacy by increasing the upward flux and sludge flotation. This is consistent with the idea that higher current densities cause an increase in the amount of aluminum dissolution, which causes more precipitation and aids in the removal of more phenol. 95.4% was the greatest removal efficiency achieved during the electrocoagulation procedure, with an electrolysis period and current density of 30 min and 100 A/m², respectively.

6.3.1.3 Color Degradation

The breakdown of ozone is directly influenced by the pH of the effluent. When ozone breaks down at higher pH levels, hydroxyl radicals are produced, and molecular ozone acts as the major oxidant at lower pH levels. Direct ozone exposure favors the degree of de-colorization at lower pH. As molecular ozone (O_3) is more prevalent at low pH than $OH^•$ radicals (which are more potent at high pH), less ozone is squandered on scavenging processes in the solution. Inorganic and organic molecules in solution, non-colored saturation bonds, and indirect $OH^•$ radical-based processes all contribute to ozone waste at high pH levels. As a result, the solution's decolorization effectiveness rises in an acidic environment (Preethi et al., 2009). Additionally, the BOT effluent's reddish hue is primarily due to the chromophore groups found in lignin. Under acidic conditions, more molecular ozone can be produced by speeding up the ozone formation process. As a result, the molecular ozone causes the chromophore groups (often double bonds) to disintegrate, which results in a bond cleavage and the loss of the capacity to absorb visible light, which causes the sample to become discolored (Adams & Gorg, 2002). As a result, the color drops from an initial content of 2450 Hazen to a final content of 250 Hazen at an optimal ozone production rate of 1.33 mg/s and treatment period of 40 min. The decolorization efficacy improves from 75.3% to 90% with an improvement in ozone production rate from 1 to 1.33 mg/s, respectively. Additionally, it was discovered that the decolorization efficiency was higher at the beginning of the experiment because molecular ozone is responsible for destroying the majority of the chromophore groups within the first 5 to 15 min of the ozonation process. However, the effectiveness of decolorization declines at higher pH because $OH^•$ radicals possess high oxidation potential and low selectivity than molecular O_3.

The color concentration further drops during electrocoagulation, from 250 Hazen (achieved at 1.33 mg/s) to 45 Hazen for an ideal electrolysis period and current density of 30 min and 100 A/m², respectively. The solution's initial pH and the amount of electricity used during electrocoagulation are both significant factors that influence how effectively the process decolorizes the sample. The contribution of various species produced throughout the experiment plays a significant role in the decolorization process. Higher initial pH inhibits the development of $Al(OH)_3$ flocs, while lower initial pH promotes hydroxy polymeric species. Even at a low pH of 3, increased reduction efficiencies are caused by the colored compounds' effective precipitation. Adsorption and precipitation are two significant interaction mechanisms that are taken into consideration, each of which is advised for a certain pH range, according to the literature (Ghosh, Medhi, & Purkait, 2009). The precipitation process is what leads to a higher decolorization efficiency, while the entrapment of polymeric constituent colloidal precipitates on the surface of $Al(OH)_3$ has a more indirect impact. The phenomenon of flocculation at pH above 6.5 can be characterized as adsorption, whereas it can be characterized as precipitation at pH between 3 and 6. When aluminum is used as a sacrificial anode and a low initial pH is applied, cationic monomers, viz.

$Al(OH)_2^+$ and Al^{3+} are the predominant components. The primary method by which the chromophore-containing organic molecules coagulate is the double-layer compression phenomenon. In this case, the colored chemicals included in the sample must be effectively removed using a coagulant (Al^{3+}) at a high concentration. In the pH range of 4–9, polymeric species $\left(Al_{13}O_4\left(OH\right)_{24}^{7+}\right)$ and precipitates such $Al(OH)_3$ (s) were produced (Changmai, Pasawan, & Purkait, 2019). As a result, the organic molecules' effective coagulation and precipitation are caused by the mechanisms of adsorption, charge neutralization, and enmeshment (having the chromophore groups). Additionally, as the electrocoagulation time increases, the capacity for decolorization also grows, demonstrating that extra flocs are continuously produced when the current density is applied to an aqueous solution. Hence, because of the similarity between the impacts of current density and electrolysis period, these two parameters can be merged to form a single variable, namely the amount of electrical consumption per cubic meter of wastewater. It can also be expressed as faradays per cubic meter. With an electrolysis period and a current density of 30 min and 100 A/m², respectively, the maximum removal efficiency that could be achieved during the electrocoagulation process was 82%. A simplified representation of the color change for the untreated and treated BOT wastewater is shown in Figure 6.2.

6.3.1.4 COD Removal

The reason for the lowering in COD and BOD content during ozonation is that ozone oxidizes the contaminants in the solution, thereby causing a drop in BOD and COD concentration. Both direct ozone oxidation and radical OH•radical oxidation are possible ozonation mechanisms. Because of the predominance of indirect OH• radical-based reactions at high pH, it is possible for significant amounts of ozone to be wasted by scavenging reactions with the inorganic and

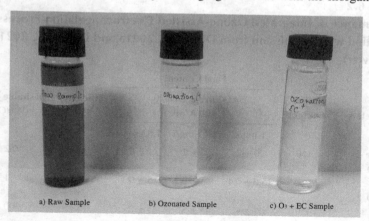

a) Raw Sample   b) Ozonated Sample   c) O₃ + EC Sample

FIGURE 6.2 Schematic illustration Of color change during (a) raw wastewater, (b) Ozonation process and (c) integrated (O_3+EC) process (Reproduced with permission from Das et al., 2021c @ Elsevier).

organic molecules in the aqueous solution. Because molecular ozone predominates at low pH, there is less ozone loss by scavenging processes compared to OH$^{\bullet}$ radicals in the solution (Adams & Gorg, 2002). As a result, under acidic circumstances, the removal efficiency of COD and BOD increases. It was found that after 40 min of ozonation, the efficacy of COD degradation improves from 58.5% to 72.2% as the ozone production rate enhances from 1 to 1.33 mg/s. However, after 40 min, there was no discernible improvement in COD elimination. As a result, it was decided that a treatment time and an ozone production rate of 40 min and 1.33 mg/s, respectively, were the ideal working parameters. It was observed that the COD concentration decreased from its initial value of 2050 mg/L to its end value of 570 mg/L.

Additionally, the electrocoagulation procedure reduces the COD content from 570 (achieved at 1.33 mg/s) to 110 mg/L at an ideal electrolysis time of 30 min and a current density of 100 A/m^2. The solution's COD content was significantly reduced as a result of the cathode's generation of OH$^-$ ions, which have exceptional absorption characteristics. Al^{3+}, a free cation, predominates in an acidic environment. The COD content is decreased during electrocoagulation by charge neutralization of colloids along with entrapment of contaminants onto the surface of Al(OH)$_3$ flocs, produced as a result of electrostatic forces. Pollutants are removed with the aid of a rise in the concentration of cationic metal ions with increasing electrolysis time (Barrera Díaz & González-Rivas, 2015). It may be said that pollutant elimination is directly related to the pace at which reaction time and current density are growing. An electrolysis period of 30 min and a current density of 100 A/m^2 resulted in a maximum removal effectiveness of 80.8% during electrocoagulation. Table 6.2 displays the degradation

TABLE 6.2
Performance of Integrated Ozone-Assisted Electrocoagulation Process (Modified with Permission from Das et al., 2021a and Das et al., 2021c @ Elsevier)

Parameters	Initial Content	Final Content			Efficiency (%)	Desirable Level (WHO)
		At Current Density 100 A/m^2				
		1 mg/s	1.11 mg/s	1.33 mg/s		
Cyanide (mg/L)	150	20 ± 2	12.5 ± 2	0.1 ± 0.05	99.8	0.2
Phenol (mg/L)	150	15.0 ± 0.4	9.5 ± 0.2	0.5 ± 0.1	99.5	1
Color (Hazen)	2450	300 ± 4	205 ± 5	45 ± 2	98.2	300
COD (mg/L)	2050	340 ± 4	250 ± 5	110 ± 3	94.7	250

efficacy of the integrated ozone-assisted electrocoagulation process for the treatment of BOT wastewater.

6.3.2 ESTIMATION OF ENERGY CONSUMPTION AND OPERATING COST

A preliminary cost study was carried out during ozonation, focusing mainly on the cost of energy usage for the entire process. It was observed that the operational and energy costs fluctuate as ozonation duration increases for an ideal ozone production rate of 1.33 mg/s. The operating cost estimate was made in accordance with the equation given below:

$$\text{Operating Cost}_{(ozonation)} = q \times Q_{energy} \qquad (6.5)$$

Here, Q_{energy} is the amount of electricity required to remove the pollutants, and the electrical energy rate is given by "q." The cost of energy usage was calculated by utilizing the electricity price determined for the state of Assam (India) in the year 2020. The amount of electrical energy used can be stated as follows (Das et al., 2021c):

$$Q_{energy} = \frac{\eta G \times G_{O_3} \times C \times t}{V_L} \qquad (6.6)$$

Here, ηG stands for the specific energy required to produce ozone (21.97 kWh/kg O_3), G_{O_3} stands for the rate of ozone production in mg/s, C stands for the unit conversion factor, t stands for the duration of ozonation experiment in min, as well as V_L stands for the wastewater volume in m^3. Energy costs were shown to rise from 7.13 to 27.53 US\$/m^3, with a rise in operating time, when the ozonation procedure was assumed to consume 51 W of electricity at an ozone production rate of 1 mg/s. As a result, the operating expense rises from 0.66 US\$/m^3 for the first 10 min of operation to 2.67 US\$/m^3 after 40 min. Similarly, it was discovered that when the operating duration rises, the energy cost increases from 14.71 to 58.61 US\$/m^3 for an ozone production rate of 1.33 mg/s and a power usage of 105.2 W. This results in a rise in operating costs from 1.37 US\$/m^3 for the first 10 min of operation to 5.48 US\$/m^3 at the conclusion of 40 min. Similar to this, the feasibility of the electrocoagulation method is determined by the overall cost incurred during the course of the experiment. Electricity, chemicals, sludge disposal, electrode costs, and fixed costs are included in the cost of operation in this case. To keep things simple, only the price of electrodes and the cost of energy are taken into account when calculating the running costs for the electrocoagulation process. Following is the calculation for the operating cost (Ghosh, Solanki, & Purkait, 2008):

$$\text{Operating Cost}_{(electrocoagulation)} = p \times Q_{electrode} + q \times Q_{energy} \qquad (6.7)$$

Here, $Q_{electrode}$ and Q_{energy} refer to the usage of electrode material and electrical energy, respectively. Further, "p" stands for the cost of electrodes, and "q" stands for the power cost for the Indian state of Assam in the year 2020. The expression for electrical energy usage is as follows:

$$Q_{energy} = \frac{I \times V \times t}{V_L} \tag{6.8}$$

Here, I stands for current (A), t stands for electrolysis time (s), V stands for voltage (V), and V_L stands for wastewater volume (m³). Following is an estimate of electrode usage based on Faraday's law (Changmai et al., 2022):

$$Q_{electrode} = \frac{MW \times I \times t}{V_L \times F \times z} \tag{6.9}$$

Here, MW stands for the molar mass of aluminum, I stands for the current (A), t stands for the electrolysis duration (s), V_L stands for the wastewater volume (m³), F stands for the Faraday constant, and z stands for the number of transported electrons. For an ideal current density of 100 A/m², the change in energy cost, electrode cost, and operating cost with electrocoagulation time was determined. Because of the increased energy usage and improved metal dissolution from anodes, the electrode cost and energy consumption increase as the current density increases. Hence, the cost of electrode consumption was found to increase from 0.00061 to 0.0018 US$/m³, while the cost of energy usage was raised from 1.57 to 4.70 US$/m³. After 30 min of the experiment, the total operational cost was enhanced from 0.157 to 0.443 US$/m³ with increasing current density from 50 to 150 A/m², respectively. Consequently, the expression representing the overall operational cost of the integrated process after 70 min (ozone-assisted electrocoagulation) is as follows:

$$\text{Operating Cost}_{(total)} = \text{Operating Cost}_{(ozonation)} + \text{Operating Cost}_{(electrocoagulation)}$$
$$= 5.822 \text{ US\$/m}^3$$

6.4 CONCLUSION

In order to treat color, phenol, cyanide, and COD from actual BOT water from the Tata steel mill, a hybrid ozonation-aided electrocoagulation technique was established in this study. An improvement in the rate of oxidation of the contaminants in the BOT water is caused by increased ozone formation rate and current density. Due to the generation of acidic by-products (inorganic acids and organic anions) throughout the ozonation procedure, the original pH was found to fall from 7.72 to 3.0. On the other hand, after the subsequent electrocoagulation procedure, the pH increased to 8.2. The outcomes demonstrated that the ozonation process alone was adequate to reduce the color below its permitted level for an optimal ozone production rate of 1.33 mg/s. Nonetheless, the combined method of ozonation-assisted electrocoagulation was needed to reduce the concentration of other pollutants, including phenol, cyanide, and COD up to 0.5 mg/L, 0.1 mg/L, and 110 mg/L, respectively. The anticipated preliminary cost for the combined ozone-assisted electrocoagulation process was 5.822 US$m^3. The hybrid process's treatment cost is comparable to and even less than that of the published literature. In terms of contaminant degradation and cost-effectiveness, the hybrid process was shown to be quite effective. Therefore, this approach might be effective and helpful for treating BOT effluent from the steel industrial sector. Nevertheless, a pilot plant study needs to be conducted before building a plant on an industrial scale.

REFERENCES

Adams, C. D., & Gorg, S. (2002). Effect of pH and gas-phase ozone concentration on the decolorization of common textile dyes. *Journal of Environmental Engineering*, *128*(3), 293–298. https://doi.org/10.1061/(asce)0733-9372(2002)128:3(293)

Barrera Díaz, C. E., & González-Rivas, N. (2015). The use of Al, Cu, and Fe in an integrated electrocoagulation-ozonation process. *Journal of Chemistry*, *2015*(1), 1–7. https://doi.org/10.1155/2015/158675

Barriga-Ordonez, F., Nava-Alonso, F., & Uribe-Salas, A. (2006). Cyanide oxidation by ozone in a steady-state flow bubble column. *Minerals Engineering*, *19*(2), 117–122. https://doi.org/10.1016/j.mineng.2005.09.001

Changmai, M., Das, P. P., Mondal, P., Pasawan, M., Sinha, A., Biswas, P., & Purkait, M. K. (2022). Hybrid electrocoagulation–microfiltration technique for treatment of nanofiltration rejected steel industry effluent. *International Journal of Environmental Analytical Chemistry*, *102*(1), 62–83. https://doi.org/10.1080/03067319.2020.1715381

Changmai, M., Pasawan, M., & Purkait, M. K. (2019). Treatment of oily wastewater from drilling site using electrocoagulation followed by microfiltration. *Separation and Purification Technology*, *210*(August 2018), 463–472. https://doi.org/10.1016/j.seppur.2018.08.007

Das, P. P., Anweshan, Mondal, P., Sinha, A., Biswas, P., Sarkar, S., & Purkait, M. K. (2021a). Integrated ozonation assisted electrocoagulation process for the removal of cyanide from steel industry wastewater. *Chemosphere*, *263*, 128370. https://doi.org/10.1016/j.chemosphere.2020.128370

Das, P. P., Anweshan, & Purkait, M. K. (2021b). Treatment of cold rolling mill (CRM) effluent of steel industry. *Separation and Purification Technology*, *274*, 119083. https://doi.org/10.1016/j.seppur.2021.119083

Das, P. P., Mondal, P., Anweshan, Sinha, A., Biswas, P., Sarkar, S., & Purkait, M. K. (2021c). Treatment of steel plant generated biological oxidation treated (BOT) wastewater by hybrid process. *Separation and Purification Technology*, *258*, 118013. https://doi.org/10.1016/j.seppur.2020.118013

Das, P. P., Mondal, P., & Purkait, M. K. (2022a). Recent advances in synthesis of iron nanoparticles via green route and their application in biofuel production, *Green Nano Solution for Bioenergy Production Enhancement* (pp. 79–104). https://doi.org/10.1007/978-981-16-9356-4_4

Das, P. P., Sharma, M., & Purkait, M. K. (2022b). Recent progress on electrocoagulation process for wastewater treatment: A review. *Separation and Purification Technology*, *292*(April), 121058. https://doi.org/10.1016/j.seppur.2022.121058

Dhadge, V. L., Medhi, C. R., Changmai, M., & Purkait, M. K. (2018). House hold unit for the treatment of fluoride, iron, arsenic and microorganism contaminated drinking water. *Chemosphere*, *199*, 728–736. https://doi.org/10.1016/j.chemosphere.2018.02.087

Do, J. S., & Chen, M. L. (1994). Decolourization of dye-containing solutions by electrocoagulation. *Journal of Applied Electrochemistry*, *24*(8), 785–790. https://doi.org/10.1007/BF00578095

Dong, G., Chen, B., Liu, B., Hounjet, L. J., Cao, Y., Stoyanov, S. R., & Zhang, B. (2022). Advanced oxidation processes in microreactors for water and wastewater treatment: Development, challenges, and opportunities. *Water Research*, *211*, 118047. https://doi.org/10.1016/j.watres.2022.118047

Ghosh, D., Medhi, C. R., & Purkait, M. K. (2008). Treatment of fluoride containing drinking water by electrocoagulation using monopolar and bipolar electrode connections. *Chemosphere*, *73*(9), 1393–1400. https://doi.org/10.1016/j.chemosphere.2008.08.041

Ghosh, D., Medhi, C. R., & Purkait, M. K. (2009). Treatment of drinking water containing iron using electrocoagulation. *International Journal of Environmental Engineering*, *2*(1–3), 212–227. https://doi.org/10.1504/IJEE.2010.029829

Ghosh, D., Medhi, C. R., & Purkait, M. K. (2011). Techno-economic analysis for the electrocoagulation of fluoride-contaminated drinking water. *Toxicological and Environmental Chemistry*, *93*(3), 424–437. https://doi.org/10.1080/02772248.2010.542158

Ghosh, D., Solanki, H., & Purkait, M. K. (2008). Removal of Fe(II) from tap water by electrocoagulation technique. *Journal of Hazardous Materials*, *155*(1–2), 135–143. https://doi.org/10.1016/j.jhazmat.2007.11.042

Kariim, I., Abdulkareem, A. S., Tijani, J. O., & Abubakre, O. K. (2020). Development of MWCNTs/TiO2 nanoadsorbent for simultaneous removal of phenol and cyanide from refinery wastewater. *Scientific African*, *10*, e00593. https://doi.org/10.1016/j.sciaf.2020.e00593

Ma, D., Yi, H., Lai, C., Liu, X., Huo, X., An, Z., & Yang, L. (2021). Critical review of advanced oxidation processes in organic wastewater treatment. *Chemosphere*, *275*, 130104. https://doi.org/10.1016/j.chemosphere.2021.130104

Mondal, P., & Purkait, M. K. (2017). Green synthesized iron nanoparticle-embedded pH-responsive PVDF-co-HFP membranes: Optimization study for NPs preparation and nitrobenzene reduction. *Separation Science and Technology*, *52*(14), 2338–2355.

Mondal, P., Samanta, N. S., Meghnani, V., & Purkait, M. K. (2019). Selective glucose permeability in presence of various salts through tunable pore size of pH responsive PVDF-co-HFP membrane. *Separation and Purification Technology*, *221*, 249–260.

Mukherjee, R., Mondal, M., Sinha, A., Sarkar, S., & De, S. (2016). Application of nanofiltration membrane for treatment of chloride rich steel plant effluent. *Journal of Environmental Chemical Engineering*, *4*(1), 1–9. https://doi.org/10.1016/j.jece.2015.10.038

Patel, S., Majumder, S. K., Das, P., & Ghosh, P. (2019). Ozone microbubble-aided intensification of degradation of naproxen in a plant prototype. *Journal of Environmental Chemical Engineering*, *7*(3), 103102. https://doi.org/10.1016/j.jece.2019.103102

Preethi, V., Parama Kalyani, K. S., Iyappan, K., Srinivasakannan, C., Balasubramaniam, N., & Vedaraman, N. (2009). Ozonation of tannery effluent for removal of cod and color. *Journal of Hazardous Materials*, *166*(1), 150–154. https://doi.org/10.1016/j.jhazmat.2008.11.035

Purkait, M. K., Sinha, M. K., Mondal, P., & Singh, R. (2018a). Interface Science and Technology, Elsevier, Pages 115–144.

Purkait, M. K., Sinha, M. K., Mondal, P., & Singh, R. (2018b). Ch 2 - pH-Responsive Membranes, Singh, *Interface Science and Technology*, Elsevier, Volume 25, Pages 39–66, ISBN 9780128139615.

Purkait, M. K., Sinha, M. K., Mondal, P., & Singh, R. (2018c). Chapter 4 - Photoresponsive Membranes, Editor(s): Mihir Kumar Purkait, Manish Kumar Sinha, Piyal Mondal, Randeep Singh, Interface Science and Technology, Elsevier, Volume 25, Pages 115–144, ISBN 9780128139615.

Purkait, M. K., Mondal, P., & Chang, C.-T. (2019). Treatment of Industrial Effluents: Case Studies, CRC Press, 1st Edition, ISBN 9781138393417.

Purkait, M. K., Singh, R., Mondal, P., & Haldar, D. (2020). Thermal Induced Membrane Separation Processes, Elsevier, ISBN 9780128188019.

Ramseier, M. K., & von Gunten, U. (2009). Mechanisms of phenol ozonation-kinetics of formation of primary and secondary reaction products. *Ozone: Science and Engineering*, *31*(3), 201–215. https://doi.org/10.1080/01919510902740477

Rowley, W. J., & Otto, F. D. (1980). Ozonation of cyanide with emphasis on gold mill wastewaters. *The Canadian Journal of Chemical Engineering*, *58*(5), 646–653. https://doi.org/10.1002/cjce.5450580516

Samanta, N. S., Banerjee, S., Mondal, P., Anweshan, Bora, U., & Purkait, M. K. (2021). Preparation and characterization of zeolite from waste Linz-Donawitz (LD) process slag of steel industry for removal of Fe^{3+} from drinking water. *Advanced Powder Technology*, *32*(9), 3372–3387.

Samanta, N. S., Das, P. P., Mondal, P., Changmai, M., & Purkait, M. K. (2022). Critical review on the synthesis and advancement of industrial and biomass waste-based zeolites and their applications in gas adsorption and biomedical studies. *Journal of the Indian Chemical Society*, *99*(11), 100761. https://doi.org/10.1016/j.jics.2022.100761

Saravanan, A., Deivayanai, V. C., Kumar, P. S., Rangasamy, G., Hemavathy, R. V., Harshana, T., & Alagumalai, K. (2022). A detailed review on advanced oxidation process in treatment of wastewater: Mechanism, challenges and future outlook. *Chemosphere*, *308*(P3), 136524. https://doi.org/10.1016/j.chemosphere.2022.136524

Sharma, M., Das, P. P., Sood, T., Chakraborty, A., & Purkait, M. K. (2022). Reduced graphene oxide incorporated polyvinylidene fluoride/cellulose acetate proton exchange membrane for energy extraction using microbial fuel cells. *Journal of Electroanalytical Chemistry*, *907*(November 2021), 115890. https://doi.org/10.1016/j.jelechem.2021.115890

Sharma, N. K., & Philip, L. (2016). Combined biological and photocatalytic treatment of real coke oven wastewater. *Chemical Engineering Journal*, *295*, 20–28. https://doi.org/10.1016/j.cej.2016.03.031

Yang, L. P., Hu, W. Y., Huang, H. M., & Yan, B. (2010). Degradation of high concentration phenol by ozonation in combination with ultrasonic irradiation. *Desalination and Water Treatment*, *21*(1–3), 87–95. https://doi.org/10.5004/dwt.2010.1233

7 Treatment of Steelmaking Wastewater: Case study

7.1 OVERVIEW OF STEELMAKING PROCESS WASTEWATER

Steelmaking is a critical step in the iron and steel industries. The essential of steelmaking processes include the basic oxygen furnace (BOF), electric arc furnace (EAF), and ladle furnace (LF) refining methods. Unlike other processes, the steelmaking does not generate a lot of wastewater, but it does generate a lot of solid waste in the form of slags. Wastewater is generated, however, as a result of slag crushing. Slags produced during the steelmaking process include basic oxygen furnace (BOF) slag, Linz-Donawitz (LD) slag, electric arc furnace slag, and ladle slag. Slag is a waste product of the steelmaking process and it consists of metal oxides, SiO_2, metal sulfides, and elemental metals that develop on the surface of impure molten metals (Mondal & Purkait, 2018). Metals in their pure form are known as ores because they exist in nature as oxides, sulfides, and other compounds or as a mixture of these metallic and non-metallic compounds. Reducing ores from chemical compound form to pure metal form necessitates the addition of flux elements that link with impurities to form and remove slag from molten metal. Flux minerals often utilized for this purpose include dolomite and burnt lime (calcium oxide). The slag layer that forms on the surface of high-temperature molten metal during ore melting is removed and solidified using various cooling processes. In the steel and iron manufacturing processes, massive amounts of slag and other by-products are created. Due to storage restrictions for solid waste and other environmental concerns, recycling such waste materials has become an extremely important topic for the protection of natural resources. Aside from such environmental issues, several attempts have been made in recent decades to utilize industrial wastes in industrial applications for energy and cost savings, as well as their recycling into usable by-products. In light of local and worldwide steel production, using slag as a value-added by-product or as extra energy output becomes both a requirement for environmental concerns and a potential for cost-saving in industrial applications. Industrial slags are also used in some low-value-added industries. These uses include filler material in road construction, additive in cement production, railway ballast manufacturing, glass production, thermal insulation wool production, membrane filters (Mondal & Purkait, 2017; Mondal, Samanta, Meghnani, & Purkait, 2019; Purkait, Mondal, & Chang, 2019; Purkait, Singh, Mondal, & Haldar, 2020), and wastewater treatment (Oge, Ozkan, Celik, Sabri Gok, & Cahit Karaoglanli, 2019).

 DOI: 10.1201/9781003366263-7

7.2 ENVIRONMENTAL ISSUES OF WASTE-GENERATED SLAG

Some researchers have noticed that releasing metals and components from slag may cause environmental issues such as air, water, and soil pollution. The inhalation of small slag particles (less than 10 m) from these dry slag dumps poses a toxicological danger to the human body. Leachates pollute soil, groundwater, and surface water bodies. To substantiate this statement, the leaching parameters were investigated by conducting a toxicity characteristic leaching process test on iron and steel slags in the United States. It was discovered that trace elements, such as nickel, chromium, lead, zinc, titanium, and vanadium, may be present. The slag contains more chromium, but the quantities in the leachate are modest because the chromium ions are bonded inside stable crystalline phases. None of the metals in the leachates have been found to surpass the threshold concentrations. Because of the existence of free lime, Linz-Donawitz (LD) slag can be quite alkaline. Leachate runoff from exposed steel slag heaps, embankment fills, or granular bases can have high pH values, which can have an impact on the nearby aquatic ecosystem. Because of their high melting temperatures, organic molecules and compounds do not exist in slag. The disposal of slags results in a lack of land and other issues (Chand, Paul, & Kumar, 2016).

7.3 UTILIZATION OF STEEL SLAG

Slags from the steel industry are effectively used as a construction material for their excellent technical qualities. Volume stability and heavy metal leaching, such as Cr and V, are potential issues with using steelmaking slags. The increased free CaO induces hydration and structural cracking. Steel slag powders are also trademarked for wastewater treatment, preventing environmental damage caused by slags and lowering the cost of wastewater treatment. Slag compositions, in general, determine slag application. Higher levels of free lime and magnesia are required for hydration fertilizer production. In road and canal construction and concrete buildings, care must be taken to avoid free lime and magnesia formation in the slag. Steel slag is an excellent material for the reclamation of acidic mine sites. The application rates required to neutralize the overall potential acidity of mine soil are large, and reapplying lime may not be physically or economically possible. All three steel slags generated were found to be as effective as limestone in neutralizing mining waste. Steel LD slag can also be used to make fertilizer for agricultural purposes (Changmai et al., 2022; Motz & Geiseler, 2001). Fertilizer components in steel slag include oxides of calcium, magnesium and silica, phosphorous, and calcium that have been utilized for a wide variety of agricultural uses. Its alkaline nature neutralizes soil acidity. Because of their significant hardness and cementing qualities, they are utilised for floor preparation and road construction. Many steel mills sell more than half of their LD slag for building and ground filling. Indian Railways employ LD slag as a railway ballast material since it has shown to be a good material. Because lime and magnesia are present, LD slag absorbs moisture and CO_2 from the atmosphere to generate hydroxides and carbonates, causing volume expansion or swelling in road or building materials (Samanta et al., 2021). This issue can be solved by weathering the slag for 6–9 months before use to allow free lime to hydrate. One of the most recent trends in industrial slag reutilization is its recycling as a green source in ceramic tile manufacture. The chemical composition of electric arc furnace (EAF) slag, in particular, makes it an alternate source for ceramic tile fabrication. EAF slag has a similar composition to ceramic tile raw materials such as silica, feldspar, and clay. EAF slag was reported to be a non-hazardous green raw material for ceramic tile manufacturing as a result of tests such as loss of ignition, flow button, and phase analysis, in addition to its composition, which is pretty similar to proper raw materials for ceramic tile production. Furthermore, much study has been conducted on using slag as a treatment media in the form of adsorbents and filters for wastewater treatment. More emphasis is still being placed on the comparative evaluation of slags of various compositions, as well as the examination of their suitability in high value added surface engineering and automotive applications.

7.4   CASE STUDY

This section discusses the practical setup for using steelmaking waste (slag) as a useful resource. It deals with the experimental setup of the process along with the process parameters optimization. The compatibility of the reuse strategy was studied and discussed in order to give a better insight into the experimental variations.

7.4.1   EXPERIMENTAL

7.4.1.1   Raw Materials for Membrane Fabrication

Basic oxygen furnace slag or LD slag and cold roll mill (CRM) effluent are provided by TATA Steel Limited, Jamshedpur, India. Merck India supplied sodium metasilicate, boric acid, sodium carbonate, alumina, quartz, and poly aluminum chloride (PAC). Loba Chemie, Mumbai, India, provided the acetic acid (99%) and kaolin. Along with BOF slag, some precursors such as sodium metasilicate, boric acid, sodium carbonate, alumina, and quartz are used in the preparation of ceramic membranes, which aid in the formation of a membrane with properties such as binding properties, mechanical strength, pore formation, plasticity, and thermal stability.

7.4.1.2   Scheme of Membrane Preparation

The membrane fabrication process is depicted in Figure 7.1. The slag was sieved through a 45 μm standard mesh. The sieved slag was washed with water and dried for 12 hours at 100°C. The washed raw slag, modified slag, and other precursors were thoroughly combined in a ball mill at a speed of 60 rpm for an hour. To develop a disc-shaped structure, the mixture was uniaxially pressed for 2 min at a pressure of 100 kg/cm^2 in a stainless steel mold. The structure was

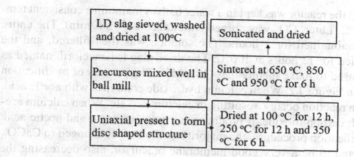

FIGURE 7.1 Scheme for membrane fabrication (Reproduced with Permission from Deepti et al., 2020, Copyright © Elsevier).

then dried at 100°C for 12 hours and then at 250°C for another 12 hours. An intermediate temperature of 350°C was introduced for 6 hours before sintering at three distinct temperatures of 650°C, 850°C, and 950°C. Sintering was performed at a rate of 2°C/min. The membranes were polished with silicon carbide abrasive, then cleaned with millipore water in an ultrasonication bath for 2 hours and dried for 3 hours. The obtained membrane had a diameter of 51.5 mm and a thickness of 5 mm, and it was tested for morphological features as well as permeability. According to the composition, CaO accounts for over 50% of the LD slag. According to reports, excessive free CaO causes hydration and cracking in structures (Shokri, Ahsan, Liu, & Muslim, 2015). Furthermore, since CaO combines with water to generate f-CaO, there is a chance of CaO leaching into the permeate after filtration. This viewpoint incorporated a modification phase to address the aforementioned concerns. Figure 7.2 depicts the modification procedure of LD slag. Washed slag was treated with acetic acid with a stirring speed

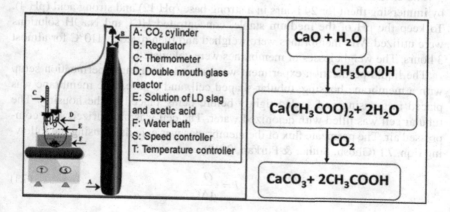

FIGURE 7.2 Modification of LD slag (a) Experimental setup and (b) Scheme of modification (Reproduced with Permission from Deepti et al., 2020, Copyright © Elsevier).

of 500 rpm, and the reactor was kept in a water bath to maintain a consistent temperature of 40°C. Later, CO_2 was purged simultaneously (2 L/min). The entire reaction was maintained for 2 hours. The solution was then filtered, and the residue was dried for 12 hours at 100°C. The dried slag is henceforth named as modified slag used to make membranes (M1, M2). The process of modification is also shown in Figure 7.2. When calcium hydroxide combines with acetic acid, a neutralization reaction occurs, resulting in calcium acetate. When calcium acetate reacts with carbon dioxide, it produces calcium carbonate and acetic acid. With this modification procedure, the majority of the CaO is changed to $CaCO_3$, which is reported to be a very good membrane precursor, also decreasing the negative effects of CaO.

7.4.1.3 Characterization Techniques

The precursors and membranes were thoroughly characterized. X-ray fluorescence was used to analyze the composition of LD slag (XRF, make: PANalytical; model: Axios). A particle size analyzer (Delsa Nano, Beckman Coulter, Switzerland) was used to examine the size distribution of all raw components and the membrane mixture. Thermogravimetric analysis was used to perform weight changes of slag and membrane mixture from 30°C to 1000°C. The pore size and any irregularities in the membrane were examined using field emission scanning electron microscopy (FESEM, make: Zeiss; model: Sigma 300). To avoid charging the specimen, a gold sputter coating was added to it prior to analysis. The average pore size of the membranes was estimated using Image J software from FESEM micrographs (Changmai & Purkait, 2018). A permeation experiment was carried out to assess the performance of membranes in separation applications. The bulk porosity of the produced membranes was assessed using the Archimedes method with water as the wetting liquid (Changmai, Pasawan, & Purkait, 2019).

The chemical durability of the produced ceramic membranes was evaluated by immersing them for 24 hours in a strong base (pH 12) and strong acid (pH 4). To keep the pH of the medium stable, concentrated HCl and NaOH solutions were utilized. Wet membranes were weighed and then dried at 110°C for almost 3 hours. The weight losses of membranes were calculated.

The liquid permeation experiment was carried out using a permeation setup with a membrane housing, tubular-shaped cell, and a base. The membrane was placed in a casing and sealed tightly before being placed in the housing. The tubular cell was filled with deionized water. The cell was pressurized with compressed air. The permeate flux of the membranes was calculated using the following Eqn. 7.1 (Ghosh, Sinha, & Purkait, 2013).

$$J = \frac{Q}{A\Delta t} \tag{7.1}$$

where Q, A, and Δt are volume of water permeated (m^3), effective membrane area (m^2), and sampling time (h), respectively.

7.4.1.4 Performance of Fabricated Membranes

One of the highest generated wastewater from steel industry, i.e., cold roll mill wastewater, was treated utilizing a hybrid method that included coagulation, flocculation, and microfiltration. This study utilized poly aluminum sulfate (PAC) as a coagulant. Coagulation-flocculation studies were performed with a jar-test setup outfitted with jars (1 L, 6 nos) with rectangular blades. The experiment was carried out in 2 stages. The coagulant dosage ranged from 30 mg/L to 500 mg/L. The following operational conditions were chosen: (rapid mixing speed = 120 rpm; gradual mixing speed = 40 rpm; settling time = 30 min). The coagulated water was then filtered by the fabricated membrane.

7.4.2 RESULTS AND DISCUSSION

This section deals with the scientific explanations for the results obtained during experimental runs. The discussions are based on the membrane characteristic feature and performance of the fabricated membrane. The below section explains the above phenomenon in a descriptive manner.

The raw materials were subjected to particle size distribution analysis before sintering. Raw materials utilized in manufacturing ceramic membranes had a wide variety of particle size distribution. After sintering, the particles help the membrane achieve a well-compacted structure with homogeneous porosity by filling in the blank spaces between the larger particles. Modified slag was observed to have the smallest size of 1.6 μm, while quartz had the largest size of 5.6 μm. Boric acid, alumina, sodium carbonate, and sodium meta silicate all had diameters of 3.8 μm, 3.5 μm, 2.3 μm, and 2.1 μm, respectively. The average particle size of the M1 and M2 mixtures are 3.5 μm and 2 μm, respectively.

Thermogravimetric analysis was performed to evaluate the temperature behavior of slag and membrane composition. It was observed that the membrane combination lost weight in two places, one for water loss and the other for calcium carbonate decomposition. No change in weight loss at 820°C implies that the minimum sintering temperature for the fabrication of the membrane must be more than 820°C. An endothermic peak around 420°C was also found, which corresponds to the dehydration of magnesium and iron hydroxides. The weight drop after 600°C is either dehydroxylation of calcium hydroxide to calcium oxide, partial breakdown of silicates, or carbonate decomposition with CO_2 release (Navarro, Díaz, & Villa-García, 2010). Field emission scanning electron microscopy was used to analyze surface morphology, such as pore distribution, shape, and size. The FESEM images of the fabricated membranes M1 and M2 at 650°C, 850°C, and 950°C are shown in Figures 7.3a and 7.3b. Within the sintering temperatures, the surface of all membranes is significantly compacted. Furthermore, pore size and pore diameter grow with increasing sintering temperature. Compared to the other membranes, the 650°C membranes exhibited far fewer pores. While the membranes sintered at 850°C and 950°C had very porous structures, this is due to the fact that sintering temperatures above 650°C promote grain growth, which leads to larger pores. Furthermore,

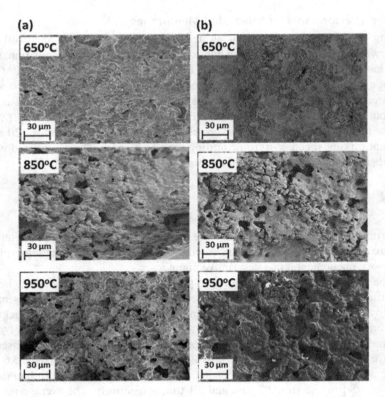

FIGURE 7.3 FESEM images of ceramic membranes at different temperatures. (a) M1 and (b) M2 (Reproduced with Permission from Deepti et al., 2020, Copyright © Elsevier).

M1 is far more porous than M2. FESEM images confirm that the sintering temperature for a porous structured membrane should be higher than 650°C, which is consistent with the TGA data. It can also be seen that the fabricated membranes are flaw-free, with an average pore size of 10 μm (within the microfiltration range) (Purkait, 2018).

The pore size distribution (PSD) of all membranes was determined by ImageJ software and FESEM images (Singh, Yadav, & Purkait, 2019). FESEM images (4 nos) of each membrane were collected for sampling and analyzed using the aforesaid software. The average pore diameter from FESEM analysis of the membrane was calculated by Eqn. (7.2), considering pores to be cylindrical (Ghosh et al., 2013).

$$d_s = \left[\frac{\sum\limits_{i=1}^{n} n_i d_i^2}{\sum\limits_{i=1}^{n} n_i} \right]^{0.5} \qquad (7.2)$$

where d_s, d_i, n_i, and n denote the average pore diameter of the area, the diameter of the ith pore, the number of pores with diameter di, and the total number of pores estimated from FESEM images, respectively. For 650°C, 850°C, and 950°C, the average pore diameters of M1 are 5.4 µm, 6.7 µm, and 8.4 µm, and those of M2 are 3.5 µm, 7.6 µm, and 8.4 µm, respectively. The prepared membrane has a wide pore size variation within the microfiltration range. It was also observed that as the temperature rises, the pore size distribution broadens through the formation of large pores, apparently eliminating smaller pores (Nandi, Uppaluri, & Purkait, 2008). The bulk porosity of the membranes was determined using the following equation (Changmai et al., 2019).

$$p = \frac{(M_W - M_d)}{(M_w - M_a)} \times 100 \qquad (7.3)$$

where M_w, M_d, and M_a represent saturated mass, dry mass, and the mass of the membrane in its dipping condition in water. As the sintering temperature for the M1 membrane was increased from 650°C to 950°C, the porosity increased from 46% to 63%. Similarly, with the same temperature shift, the porosity of the M2 membrane increased from 53% to 71%. The increase in porosity was caused by volatile elements leaving the surface during sintering, making it porous. The rise in temperature caused the pores to open (Vasanth, Pugazhenthi, & Uppaluri, 2011).

Water permeation tests were performed on the manufactured membranes M1 and M2 sintered at 650°C, 800°C, and 950°C to measure the water flux of the membrane in batch mode operation using deionized water. Throughout the experiment, a transmembrane pressure of 101 kPa was maintained. Before conducting the flux studies, each membrane was compressed to 293 kPa to unblock all pores. The flow was seen to be very high at first, then progressively dropped and reached a steady state after 26 min of operation. Furthermore, both M1 and M2 exhibit an increasing trend of flux with increasing sintering temperature due to an increase in porosity with temperature rise. At the beginning, flux was, 430 L/(m².h), 705 L/(m².h) and 818 L/(m².h) and reached steady state value of 161 L/(m².h), 252 L/(m².h) and 340 L/(m².h) for M1 sintered at 650°C, 850°C and 950°C, respectively.

M2 had an almost similar flux pattern. For membranes sintered at 650°C, 850°C, and 950°C, the initial flux was 410 L/(m².h), 635 L/(m².h), and 705 L/(m².h), respectively, and achieved steady state values of 135 L/(m².h), 206 L/(m².h), and 320 L/(m².h). It was found that M1 and M2 had practically identical flux levels. This variance in flux levels was also caused by the porosity change stated in the prior section.

The weight loss of a ceramic membrane in acidic and basic mediums is investigated. It was observed that as the temperature was raised, the weight loss for both membranes dropped. The weight loss % for both membranes was higher during acid treatment than base treatment. M1 loses more weight in both acid and basic medium than M2. At sintering temperatures of 650°C and 950°C, the percentage weight loss following acid treatment was determined to be between 15%

and 8% for M1 and 2% to 0.5% for M2. Similarly, after base treatment, weight loss decreased from 14% to 8% and 0.6% to 0.08% for M1 and M2, respectively, at 650°C and 950°C. Ceramic materials are stated to be resilient against strong acidic and basic media at higher temperatures. As a result, this shows that the prepared membrane possesses ceramic characteristics. The percentage weight loss in acidic media is greater due to oxidation of the material, such as iron, which is present in the modified slag, and effervescence, which happens when any carbonates react with acid (Elomari et al., 2016). Since M1 contains sodium carbonate, its weight loss is greater than that of M2. As a result of the presence of sodium carbonate in M1, higher weight loss is justified.

The mechanical strength of the fabricated microfiltration membranes was determined using the three-point bending method. It was found that M1 had a flexural strength of around 8 MPa at 650°C, a slight increase at 850°C, and remained essentially constant (9MPa) at 950°C. A similar pattern was observed in the case of M2. Flexural strength was 9 MPa at 650°C and increased to 10 MPa at 850 and 950°C. When compared to M1, M2 had more mechanical strength. The increase in mechanical strength as sintering temperature rises is mainly due to grain growth in the membrane, which eventually leads to densification (Purkait, 2018).

7.4.3 PERFORMANCE OF MEMBRANE

The efficacy of coagulation-flocculation and microfiltration was independently evaluated by water quality parameters in the treated water, specifically total dissolved solids (TDS), chemical oxygen demand (COD), chromium, and iron. A preliminary investigation was carried out in order to establish an estimated range of coagulant dosage. As a result, a wide range of PAC concentrations (30–500 mg/L) were evaluated. No significant changes were detected at the lowest dosage of 30 mg/L. However, a minor difference was observed at 50 mg/L. When the dosage reached 150 mg/L, the formation of floc was seen, and at 350 mg/L, complete floc development and settling were observed.

The particle size distribution (PSD) of the flocs formed during coagulation-flocculation was determined using a particle size analyzer. The majority of the particles lay in the range of 10 μm. When the pore size distribution of the membrane is compared to the PSD of the flocs produced after coagulation-flocculation, it is inferred that the membranes (particularly M2) are more applicable for the separation of flocs from coagulated water.

The coagulated water was then passed through membrane filter M2. The permeation experiments were carried out in the manner described in the previous section. The steady flux after 25 minutes of running was 231 L/(m^2.h), while the pure water flux was 320 L/(m^2.h) (m^2.h). The decreasing trend in flux was caused by the deposition of flocs on the membrane's surface, which blocked the active pores available for microfiltration. The membrane surface was revealed to be covered in a layer of floc. However, cleaning the membrane can recover more than 98% of the flux.

The EPA allowable limit for surface water was used to compare the quality of raw CRM wastewater and processed water. It was observed that the results obtained by the hybrid process of coagulation-flocculation followed by microfiltration showed a significant reduction in all parameters. The chromium value dropped from 2.26 mg/L to 0.035 mg/L, and the iron value dropped from 5.7 mg/L to 0.51 mg/L. Turbidity increased from 12.30 NTU to 16.40 NTU due to floc development but dropped to 0.9 NTU after microfiltration. It can also be noticed that all of the water quality is within the surface water allowed limits (Epa, 2017).

7.4.4 Cost Analysis

A membrane cost study is a crucial element for validating the overall feasibility of the membrane process. Polymeric membrane for industrial use cost between 50 and 200 US\$/m² (Ghosh et al., 2013). Inorganic membranes, on the other hand, are considered to be more expensive than polymeric membranes, which cost between 500 and 1000 US\$/m² (Nandi et al., 2008). The raw material cost for the present membrane is estimated to be 32.55 US\$/m² for M1 and 55.7 US\$/m² for M2 membranes, respectively. The overall cost, including raw materials and manufacturing costs, was projected to be roughly 100 US\$/m² for M1 and 125 US\$/m² for M2 membranes, respectively. As a result, the cost of the LD slag-based membrane is very similar to that of the polymeric membrane, confirming the economic utilization of such solid waste generated in the steel industry.

7.5 SUMMARY

The steelmaking process does not generate a large volume of wastewater. However, waste in the form of slag is massively produced. Slag management in a sustainable manner is a major issue that the industry faces. However, there are numerous technologies in use as well as emerging for the use of slag in various applications, namely, construction, water treatment, and road construction. The iron and steel industry is primarily concerned with increasing the recycling of slags on a daily basis in order to save energy and natural resources and, eventually, enhance production. Newer technologies, as well as improvements to existing ones, have been researched and developed in order to meet the ambitious goal of eliminating waste in the upcoming years. The efficient use of BOF slags results in a high-value-added product that improves steel plants. Also, sustainable use of slag helps to conserve natural resources. So, the use of LD slag in various industries has not only economic but also ecological benefits, as it has led to resource conservation and thus solved disposal concerns. The case study highlights the importance of LD slag and its application in the production of microfiltration membranes that may be utilized to treat wastewater from the steel industry or any other effluent. Using slag with necessary modifications for different applications is not only an appealing option for confirming ceramic membrane commercialization but also a viable option for avoiding solid waste dumping into nature.

REFERENCES

Chand, S., Paul, B., & Kumar, M. (2016). Sustainable approaches for LD slag waste management in steel industries: A review. *Metallurgist, 60*(1–2), 116–128. https://doi.org/10.1007/s11015-016-0261-3

Changmai, M., Pasawan, M., & Purkait, M. K. (2019). Treatment of oily wastewater from drilling site using electrocoagulation followed by microfiltration. *Separation and Purification Technology, 210*(July 2018), 463–472. https://doi.org/10.1016/j.seppur.2018.08.007

Changmai, M., & Purkait, M. K. (2018). Detailed study of temperature-responsive composite membranes prepared by dip coating poly (2-ethyl-2-oxazoline) onto a ceramic membrane. *Ceramics International, 44*(1), 959–968. https://doi.org/10.1016/j.ceramint.2017.10.029

Changmai, M., Das, P. P., Mondal, P., Pasawan, M., Sinha, A., Biswas, P., Sarkar, S., & Purkait, M. K. (2022). Hybrid electrocoagulation–microfiltration technique for treatment of nanofiltration rejected steel industry effluent. *International Journal of Environmental Analytical Chemistry, 102*(1), 62–83.

Deepti., Sinha, A., Biswas, P., Sarkar, S., Bora, U., & Purkait, M. K. (2020). Separation of chloride and sulphate ions from nanofiltration rejected wastewater of steel industry. *Journal of Water Process Engineering, 33*(December 2019), 101108. https://doi.org/10.1016/j.jwpe.2019.101108

Elomari, H., Achiou, B., Ouammou, M., Albizane, A., Bennazha, J., Alami Younssi, S., & Elamrani, I. (2016). Elaboration and characterization of flat membrane supports from Moroccan clays. Application for the treatment of wastewater. *Desalination and Water Treatment, 57*(43), 20298–20306. https://doi.org/10.1080/19443994.2015.1110722

Epa, U. S. (2017). *Water Quality Standards Handbook Chapter 3 : Water Quality Criteria.*

Ghosh, D., Sinha, M. K., & Purkait, M. K. (2013). A comparative analysis of low-cost ceramic membrane preparation for effective fluoride removal using hybrid technique. *Desalination, 327*, 2–13. https://doi.org/10.1016/j.desal.2013.08.003

Mondal, P., & Purkait, M. K. (2017). Green synthesized iron nanoparticle-embedded pH-responsive PVDF-co-HFP membranes: Optimization study for NPs preparation and nitrobenzene reduction. *Separation Science and Technology, 52*(14), 2338–2355.

Mondal, P., Samanta, N. S., Meghnani, V., & Purkait, M. K. (2019). Selective glucose permeability in presence of various salts through tunable pore size of pH responsive PVDF-co-HFP membrane. *Separation and Purification Technology, 221*, 249–260.

Motz, H., & Geiseler, J. (2001). Products of steel slags an opportunity to save natural resources. *Waste Management, 21*(3), 285–293. https://doi.org/10.1016/S0956-053X(00)00102-1

Nandi, B. K., Uppaluri, R., & Purkait, M. K. (2008). Preparation and characterization of low cost ceramic membranes for micro-filtration applications. *Applied Clay Science, 42*(1–2), 102–110. https://doi.org/10.1016/j.clay.2007.12.001

Navarro, C., Díaz, M., & Villa-García, M. A. (2010). Physico-chemical characterization of steel slag. Study of its behavior under simulated environmental conditions. *Environmental Science and Technology, 44*(14), 5383–5388. https://doi.org/10.1021/es100690b

Oge, M., Ozkan, D., Celik, M. B., Sabri Gok, M., & Cahit Karaoglanli, A. (2019). An overview of utilization of blast furnace and steelmaking slag in various applications. *Materials Today: Proceedings, 11*, 516–525. https://doi.org/10.1016/j.matpr.2019.01.023

Purkait, M. K., R. Singh. (2018). Membrane Technology in Separation Science. CRC Press, ISBN 9781315229263

Purkait, M. K., Sinha, M. K., Mondal, P., & Singh, R. (2018a). Interface Science and Technology, Elsevier, Pages 115–144.

Purkait, M. K., Sinha, M. K., Mondal, P., & Singh, R. (2018b). Chapter 2 - pH-Responsive Membranes, Singh, Interface Science and Technology, Elsevier, Volume 25, 2018, Pages 39–66, ISBN 9780128139615.

Purkait, M. K., Mondal, P., & Chang, C.-T. (2019), Treatment of Industrial Effluents: Case Studies, CRC Press, 1st Edition, ISBN 9781138393417.

Purkait, M. K., Singh, R., Mondal, P., & Haldar, D. (2020). Thermal Induced Membrane Separation Processes, Elsevier, ISBN 9780128188019.

Samanta, N. S., Banerjee, S., Mondal, P., Anweshan, Bora, U., & Purkait, M. K. (2021). Preparation and characterization of zeolite from waste Linz-donawitz (LD) process slag of steel industry for removal of Fe^{3+} from drinking water. *Advanced Powder Technology*, 32(9), 3372–3387.

Shokri, M., Ahsan, A., Liu, H. Y., & Muslim, N. H. (2015). An overview of use of Linz-Donawitz (LD) steel slag in agriculture. *Journal of Advanced Review on Scientific Research*, 5(1), 30–41.

Singh, R., Yadav, V. S. K., & Purkait, M. K. (2019). Cu2O photocatalyst modified anti-fouling polysulfone mixed matrix membrane for ultrafiltration of protein and visible light driven photocatalytic pharmaceutical removal. *Separation and Purification Technology*, 212(October 2018), 191–204. https://doi.org/10.1016/j.seppur.2018.11.029

Vasanth, D., Pugazhenthi, G., & Uppaluri, R. (2011). Fabrication and properties of low cost ceramic microfiltration membranes for separation of oil and bacteria from its solution. *Journal of Membrane Science*, 379(1–2), 154–163. https://doi.org/10.1016/j.memsci.2011.05.050

8 Treatment of Cold Rolling Wastewater
Case Study

8.1 INTRODUCTION

For more than 150 years, the global steel sector has been the engine of economic growth. A large rise in effluent discharge has occurred as a result of the steady scaling up of steel production over time. The average amount of water used by Tata Steel, Jamshedpur (India), to create 1 ton of steel is 25 m³, and the amount of wastewater produced each operating hour is 5 m³ (Mukherjee, Mondal, Sinha, Sarkar, & De, 2016). Wastewater is produced throughout several unit activities of the steelmaking process, including the cold and hot rolling mills, Linz-Donawitz (LD) converter, blast furnace, and a number of auxiliary operations. Among these different processes, the wastewater from the cold rolling mill (CRM) process is one of the most inevitable waste products, requiring additional treatment and neutralization before being discharged into the environment by enterprises. Using a number of subprocesses, including tempering, pickling, annealing, and rolling, the CRM procedure gives steel the desired hardness, thickness reduction, and finishing. The process produces a significant amount of oily wastewater, the most difficult of which to treat is the emulsion effluent from cold rolling (Das, Anweshan, & Purkait, 2021b).

CRM wastewater, which are byproducts of the steel-fabrication process, have a significant potential for carbonation due to their very alkaline nature. When discharged, CRM effluents typically contain iron, phenol, acid, alkaline, and oil and grease. Depending on the composition of the steel, additional substances might be present in the effluent, viz. different salts of zinc, chromium, nickel, and copper (Colla et al., 2017). The effluent must be treated before being released into the environment in order to meet the required water quality parameter. Some of the frequently employed treatment techniques include electrochemical catalytic oxidation (Zhang et al., 2010), biological methods (Sharma, Das, Sood, Chakraborty, & Purkait, 2021), ultrafiltration membrane separation (Mondal & Purkait, 2017; Mondal, Samanta, Meghnani, & Purkait, 2019; Purkait, Mondal, & Chang, 2019; Purkait, Singh, Mondal, & Haldar, 2020; Symons, 1971), chemical coagulation (Ashraf et al., 2016), and magnetic filtration (Oberteuffer, Wechsler, Marston, & Mcnallan, 1975). Despite the positive results, each approach has its own set of drawbacks. For instance, industrial effluents with various and large loads of contaminants may cause membranes to clog, which is referred to as membrane fouling, during membrane separations (Purkait, Sinha, Mondal, & Singh, 2018a, b, c). At pressures of more than 3 bar, the

DOI: 10.1201/9781003366263-8

membrane may be harmed. Equipment is expensive, and cleaning costs might be high (Changmai et al., 2020. Additionally, rather than filtering the particles based on their size, the magnetic filtration process filters them on the basis of their composition. Heavier metal particles as such have the potential to wash downstream, block the flow, or even travel back into it. The application of such a technology for non-ferromagnetic materials is extremely limited (Oberteuffer, Wechsler, Marston, & Mcnallan, 1975; Samanta et al., 2021). In addition, boron-doped diamond (BDD) cells, which are frequently employed in electrochemical oxidation, produce good treatment efficiency but are prohibitively expensive. The procedure as a whole may incur significant increases in capital and operating costs. The amount of space needed for effective effluent treatment may make BDD's industrial application exceedingly challenging. The chemical coagulation-flocculation method also calls for a variety of chemicals and exact pH control during the procedure. Further, this technique produces a number of secondary pollutants and huge amounts of sludge, which raises serious environmental concerns (Muzyka et al., 2019). The biological process also calls for auxiliary non-pathogenic and pathogenic microbes, viz. algae, bacteria, fungi, and yeast (Sharma, Das, Sood, Chakraborty, & Purkait, 2022), is mostly employed for the treatment of industrial wastewater.

Consequently, there is an urgent requirement for a technology that is more dependable, highly systematic, feasible, significant, cost-effective, scalable, and environment-friendly with high capacity throughput. Electrocoagulation (EC) and ozonation (O_3) procedures have both received widespread acclaim for their ability to drastically reduce the pollutant load from different industrial wastewater (Das et al., 2021a). The explanation may have something to do with their strong oxidative capacity, quick reaction time, reduced energy usage, ease of use, straightforward equipment design, and low operational cost. High concentrations of contaminants can be removed from industrial and synthetic effluents using either standalone or simultaneous ozonation and electrocoagulation processes, and both methods have shown encouraging results (Das, Anweshan, & Purkait, 2021b; Das, Sharma, & Purkait, 2022). The Tata Steel Cold Rolling Mill in Jamshedpur, India, provided the effluent that was taken into consideration here. If left untreated, the effluent contains a number of pollutants in high quantity, including iron, phenol, BOD, COD, and oil content. Here, electrocoagulation and/ or ozonation may be a potential method for the remediation of CRM wastewater due to the different advantages connected with both processes listed above.

In this context, a comparative study based on ozonation and electrocoagulation processes for treating CRM wastewater from the Tata steel industry has been extensively discussed and summarized in this chapter. The experimental design of both electrocoagulation and ozonation processes has been demonstrated. The removal of contaminants such as phenol, iron, oil and grease, and COD content at different current densities, electrode distances, and operating time has also been covered in detail. Furthermore, an investigation of the operating cost and energy usage of both processes has also been included towards the end of this chapter.

8.2 MATERIALS AND EXPERIMENTAL METHODS

8.2.1 Materials

Tata Steel Ltd. Jamshedpur (India) provided CRM wastewater for conducting the experiments. The effluent from CRM was characterized, and several water quality parameters were found. To change the pH of the solution, 0.1 (M) NaOH and 98% pure H_2SO_4 were utilized. The required chemicals (analytical reagent grade) were procured from Merck (Germany), and all the analysis and experiments were carried out without further purification. Further, acetic acid and methanol of HPLC quality were purchased from Merck. Aluminum sheets were employed to make the electrode materials for the electrocoagulation tests. To maintain their effectiveness during the operations, the electrodes were cleaned and scraped with 120-grit sandpaper (Das et al., 2021c).

8.2.2 Experimental Methods

8.2.2.1 Ozonation

An ozone generator and an oxygen concentration enhancer were used to test the efficacy of the elimination of pollutants found in CRM effluent using ozonation. An average rate of 0 to 3 mg/s is used to create ozone. The reactor was fed with ozone using a sparger to create microbubbles. The glass reactor held up to two liters of fluid, and one liter of effluent was used for each analysis. The surplus O_3 was changed into O_2 by an ozone destructor. A simplified diagram representing the ozone formation mechanism is shown in Figure 8.1. Specifically, 0.85, 1.00, and 1.12 mg/s of ozone-generating rates were used in this investigation to treat the CRM wastewater (Khuntia, Majumder, & Ghosh, 2015). For the purpose of making a comparison between electrocoagulation and ozonation procedures, all tests and analyses were carried out for 30 min at ambient temperatures and pressure. The studies carried out at an ozone production rate of 1.12 mg/s or higher were shown to greatly improve the energy usage and operational cost of the procedure. Additionally, with an ozone production rate of 1.12 mg/s, the permitted levels of all the contaminants treated were successfully reached. Extending the ozonation procedure past 30 min led to a little improvement in pollutant removal. Therefore, 1.12 mg/s and 30 min were chosen as the ideal ozone formation rate and treatment time, respectively. Therefore, 1.12 mg/s (ozone formation rate) and 30 min (operational period) were chosen as the ideal parameters for all following experiments. A photometer was utilized to

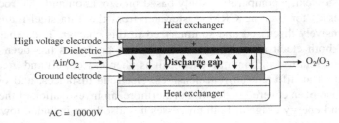

FIGURE 8.1 Schematic illustration showing the mechanism of ozone formation (Reproduced with permission from Rekhate & Srivastava, 2020 @ Elsevier).

measure the pollutant concentrations in the treated sample for both electrocoagulation and ozonation procedures after a period of every 5 min.

8.2.2.2 Electrocoagulation

The electrocoagulation procedure was carried out using an electrochemical chamber made of acrylic material with a volumetric capacity of 0.5 L. Both the cathodes and the anodes were made from aluminum sheets with dimensions of 0.07 m × 0.035 m and an acceptable electrode area of 2.45×10^{-3} m^2 (Das, Anweshan, & Purkait, 2021b). The trials carried out at current densities above 200 A m^{-2} demonstrated identical degradation efficacies regarding contaminant concentration throughout the application of three current densities, namely 200, 150, and 100 A m^{-2}, indicating that the process' removal effectiveness reaches saturation. However, current densities beneath 200 A m^{-2} were not able to eliminate all of the contaminants under their allowable levels. Therefore, the ideal current density for the experiment was determined to be 200 A m^{-2}. Similarly, a 30-min electrocoagulation period was chosen as the ideal working period because continuing the studies past this point would only marginally reduce the amount of contaminants removed. 200 A m^{-2} (current density) and 30 min (operational period) were therefore chosen as the ideal settings for all subsequent experiments. For the electrocoagulation tests in this work, aluminum was used as the electrode material. After each experiment, it was noted that the treated samples took on color when electrocoagulation was carried out using Fe-electrodes. Al-electrodes were thus employed to prevent the production of color in the final treated water. All the experiments were conducted at ambient pressure and temperature. A simplified depiction of the electrocoagulation mechanism is shown in Figure 8.2.

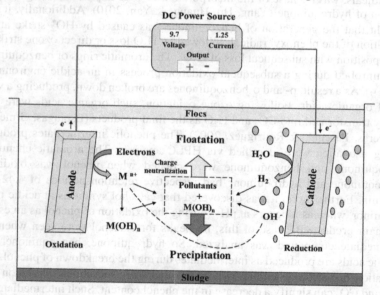

FIGURE 8.2 Schematic diagram showing the mechanism of the electrocoagulation process (Reproduced with permission from Das, Sharma, & Purkait, 2022 @ Elsevier).

8.3   RESULTS AND DISCUSSION

8.3.1   Performance of Ozonation and Electrocoagulation Process

8.3.1.1   Phenol Removal

The variation in the phenol content with different ozone production rates was determined. After 30 min of operation, it was found that the phenol degradation efficacy improved from 63.5% to 93.5% as the ozone production rate increased from 0.85 to 1.12 mg/s, respectively. The oxidation of phenol results in the generation of hydroxyl radicals. It was caused by the transfer of an electron from phenolate ion/phenol to ozone, according to the following equation (Ramseier & von Gunten, 2009):

$$PhO^- / PhOH + O_3 \rightarrow PhO^{\bullet} / PhOH^{\bullet+} + O_3^{\bullet-} \tag{8.1}$$

The direct electron transfer that occurs when ozone oxidizes phenol may be the cause of production for both the ozonide and phenoxyl radicals. As the extent of dissolved ozone in the sample rose, the rate of phenol breakdown also increased. The cause can be attributed to the reaction of ozone with both the organic and inorganic substances, which transforms them into their higher oxidation phases, along with the production of CO_2 and H_2O, respectively. Phenol oxidation produced hydroquinone at the para-position, which quickly underwent further oxidation to produce o- and p-benzoquinones. The creation of dihydroxy cyclohexadienyl radicals, which subsequently resulted in the inclusion and removal of $HO_2^{\bullet}$ and $O_2^{\bullet-}$ radicals, was because of the incorporation of •OH to phenol during the production of hydroquinone (Yang, Hu, Huang, & Yan, 2010). Additionally, it was thought that the generation of benzoquinone was caused by $HO_2^{\bullet}$ strike at the p-position of the phenoxyl radical, followed by H_2O loss or direct ozone strike at the p-position with subsequent loss of H_2O_2. The aromatic rings of benzoquinone were unrolled during a subsequent oxidation process in an acidic environment (pH = 4). As a result, p- and o-benzoquinones are broken down, producing a variety of organic acids. Following ozone oxidation, such organic acids were ultimately transformed into CO_2 and H_2O as the final products (Esplugas, Giménez, Contreras, Pascual, & Rodríguez, 2002). The phenolic intermediates produced during ozonation were identified via HPLC analysis. The aromatic chemicals hydroquinone and benzoquinone were created when phenol was oxidized. Benzoquinone and hydroquinone had respective retention periods of 4.42 and 7.92 min. Furthermore, it was discovered that catechol synthesis at acidic pH 3 was minor, whereas it occurs at pH 7 during the oxidation of phenol as an essential main product. Because of this, no peaks for catechol were seen when the intermediate compound was analyzed. So hydroquinone, benzoquinone, and organic acids are produced as intermediates during the breakdown of phenol. The formation of various phenolic by-products and their subsequent oxidation into H_2O and CO_2 can signify a decrease in the phenol content. Such intermediates do not remain in their current state for a longer time because they rapidly oxidize into H_2O and CO_2 as the finished products (Krishnan et al., 2022). As a result, it was concluded that the phenol concentration of the solution reduces as ozone oxidation increases. The concentration of phenol decreases from 15 mg/L (initial content) to 0.95 mg/L (final content) during the ideal treatment period of 30 min and ozone production rate of 1.12 mg/s.

On the other hand, after 30 min of operation, the effectiveness of phenol degradation improves from 70% to 98% when the current density is enhanced from 100 to 200 A/m^2 during the electrocoagulation process. Using aluminum electrodes, it was discovered that the effectiveness of phenol reduction increased with regard to the duration of the electrolysis. Insoluble species have been reported to be produced by the combined precipitation, coagulation, and complexation processes as a result of the chemical interaction between Al^{3+} ions and phenols (Uğurlu, Gürses, Doğar, & Yalçin, 2008). The hydrolyzed products created throughout the process adsorbed the phenolic compounds on their surface (coagulation phenomenon), and then flocs were created as a result of the physical contact of the resultant particulate constituents (flocculation phenomenon). At the cathode surface, phenol and its numerous metabolites are broken down into smaller organic molecules because of the current density. The organic molecules adhere to the surfaces of $Al(OH)_3$ and other monomeric and polymeric compounds produced in situ, where they get precipitated at the bottom or float as a result of the evolution of H_2 bubbles. During the experiment, the H_2 bubble density grew, and the bubble size simultaneously decreased, which led to an increase in the current density (Das et al., 2021c). As a result, a larger upward flux intensifies the removal efficiency, which in turn causes the floc to float more freely. The in-situ creation of $Al(OH)_3$ is also improved by high current density, which finally led to more floc precipitation and made it easier to lower the phenol content. The concentration of phenol decreased from 15 mg/L (initial content) to 0.3 mg/L (final content) as a result of the experiment, which was carried out at the ideal treatment period of 30 min and a current density of 200 A/m^2.

8.3.1.2 Iron Removal

The change in lowering the iron concentration in the wastewater with different rates of ozone formation was determined. After 30 min of operation, the iron degradation efficacy was improved from 58.5% to 86.5% as the ozone production rates were enhanced from 0.85 to 1.12 mg/s, respectively. Throughout the process, oxidation via ozone changed the Fe^{2+} ions into the Fe^{3+} ions, with the precipitation of the oxidized salts resulting in the creation of ferric hydroxide. Under acidic circumstances, Fe^{2+} rapidly oxidizes to Fe^{3+}. Fe^{2+} served as a catalyst for the hydroxyl radicals that were produced during the breakdown of ozone. Additionally, during the oxidation of Fe^{2+}, the intermediate species FeO^{2+} quickly changes into •OH. The Fe^{3+} ions were changed into $Fe(OH)_3$ based on the pH range, which then precipitated at the bottom of the reactor (El Araby, Hawash, & El Diwani, 2009). As a result, the iron degradation from the wastewater was predominantly accomplished by the formation of $Fe(OH)_3$, followed by its sedimentation. As $Fe(OH)_3$ flocs were insoluble in water under equilibrium conditions, it was thought that they would eventually settle down at the base of the ozonation set-up without engaging in any form of reaction and could therefore be easily decanted out from the treated solution (Kishimoto & Ueno, 2012). After that, the treated solution was checked for any residual Fe^{2+}. The effectiveness of removing Fe^{2+} from the solution improves as ozone solubility in water increases because more soluble iron ions will be oxidized. Thus, the concentration of iron decreases from 6.0 mg/L (initial content) to 0.8 mg/L (final content) at an ideal operational period and an ozone production rate of 30 min and 1.12 mg/s, respectively.

On the other hand, after 30 min of operation, the efficacy of iron degradation improves from 65% to 97.5% with a current density increase from 100 to 200 A/m^2 during electrocoagulation. The Fe^{2+} ion adsorption on the produced $Al(OH)_3$ flocs' surface was the main factor in iron removal by electrocoagulation. The solution pH, along with the redox potential, had a major role in controlling the state of iron species. Considering the pH to be above 4, the dissolved species of iron will be divalent in nature. Even though the bulk pH remained far below the value indicating the solubility of the product, the transfer of the produced $Fe(OH)_2$ to the liquid bulk stayed insoluble. As a result, $Fe(OH)_2$ adsorption on Al flocs was the main reason for the improvement in iron reduction efficiency (Doggaz, Attour, Le Page Mostefa, Tlili, & Lapicque, 2018). Additionally, as the current density grew, aluminum hydroxide flocs were produced in situ to a greater extent, which improved iron adsorption on their surface. As a result, the accessible binding sites were very important during iron adsorption. The formation of $Al(OH)_3$ during the hydrolysis reaction led to the production of reactive Al^{3+}-Fe^{2+} complexes, which promotes Fe^{2+} oxidation. As the pH of the solution rises, Fe^{2+} hydrolyses and forms precipitates (Ghosh, Solanki, & Purkait, 2008). Therefore, it can be deduced that as the pH rises, the iron degradation was primarily caused by the formation of $Al(OH)_3$ flocs, which improved the adsorption of $Fe(OH)_2$ on its surface and eventually precipitates at the base of the electrocoagulation chamber as reddish-brown sludge. The results of the research showed that the concentration of iron decreases from 6.0 mg/L (initial content) to 0.15 mg/L (final content) when the treatment is carried out for a period of 30 min and at a current density of 200 A/m^2.

8.3.1.3 Oil and Grease Removal

With varying ozone production rates, namely 0.85, 1.00, and 1.12 mg/s, the decrease in oil content in the CRM wastewater throughout the ozonation process was determined. After 30 min of operation, the oil reduction efficacy was found to rise from 32% to 62% as the ozone formation rate increases from 0.85 to 1.12 mg/s, respectively. Aliphatic hydrocarbons predominate among the oily pollutants in industrial wastewater. Aliphatic hydrocarbons undergo oxidation throughout the process, which leads to the generation of free radicals and the breaking of the σ-bond between H_2 and C atoms. Ketones, carboxylic acids, and alcohols were produced as intermediates as a result of these molecules' interaction with the hydroxyl ion (Lee & Coote, 2016). As soon as the ozone molecules remove the H atom from the C-H single bond system, the HO_3^* radical is created. The brief intermediates, viz. hydrogen trioxide (HOOOH) and alkyl hydrotrioxide (ROOOH), which ultimately disintegrate into H_2O and O_2, are produced when the hydrotrioxide radicals react with an alkane and again steal an H atom to create alkyl radicals (Cerkovnik, Eržen, Koller, & Plesničar, 2002). ROH, $R_2C = O$, and RCOOH, among other intermediate hydrophilic molecules, are totally mineralized via the aforementioned method. However, because of the inductive impact of the functional groups on the distribution of the molecules' electron cloud, the mineralization reactions are typically slower than the predecessors. When ozone reacts with water, the resulting •OH radicals can remove H atoms from the saturated hydrocarbons, whether in the form of cycloalkanes, branched alkanes, or straight-chained alkanes. The production of the alkyl radicals, which then reacts to generate hydrophilic constituents and finally lowers the concentration of oil in the wastewater, is the result of •OH oxidation through H atom abstraction (von Sonntag & von Gunten, 2015). As a result of reaction selectivity, this route predominates at lower pH. Thus, it was revealed that the concentration of oil decreased from 25 mg/L (initial content) to 9.5 mg/L (final content) at an ideal operational period of 30 min and an ozone formation rate of 1.12 mg/s.

Conversely, after 30 min of operation, oil reduction efficacy improves from 40% to 88% with a current density increase from 100 to 200 A/m² during electrocoagulation. It was discovered that as the current density rises from 100 to 200 A/m² during electrocoagulation, the effectiveness of oil removal improves. By raising the current imparted to all the electrodes, an increase in current density was achieved because an increase in the supplied current ultimately enhances the Al^{3+} ion dissolution from the anode. It is well known that the surface charge of the emulsified oil droplets is negatively charged, which coalesces when it is offset by the Al^{3+} ions having positive charges (Abdel-Aziz, El-Ashtoukhy, Zoromba, & Bassyouni, 2016). Additionally, the electrocoagulation process produces an amorphous $Al(OH)_4^-$ and $Al(OH)_3$ flocs, which have enhanced surface areas and aid in the oil droplet adsorption, followed by its sedimentation at the base of the electrocoagulation chamber. Further, the formation of O_2 gas at the anode and H_2 gas at the cathode contributed to the flotation of oil droplets because of their excessive upward flux, which helped to reduce the amount of oil in the wastewater (Das, Anweshan, & Purkait, 2021b). Also, it was observed that the concentration of oil decreased from 25 mg/L (initial content) to 3 mg/L (final content) at the optimal treatment duration and current density of 30 min and 200 A/m², respectively.

8.3.1.4 COD Removal

The variation in COD content in the CRM wastewater with regard to different ozone production rates was determined. After 30 min of operation, COD degradation efficacy improves from 50.5% to 70% as the ozone production rate is enhanced from 0.85 to 1.12 mg/s, respectively. Ozonation causes the pollutants in the effluent to be oxidized, which lowers both the COD and the BOD concentration. The two processes by which ozone was oxidized were radical oxidation via hydroxyl radicals, which were less selective and predominated in basic medium, or direct oxidation via molecular O_3, which was more selective and predominated under acidic medium. The scavenging action with both organic and inorganic constituents of the wastewater under alkaline circumstances may have the potential to waste a significant quantity of ozone due to the predominance of indirect hydroxyl radical-based processes. However, in acidic conditions, molecular ozone predominated, and as a result, ozone wastage through scavenging reactions in the solution was significantly lower than that of •OH radicals. As a result, at low pH, both COD and BOD content reduction efficiency rose (Adams & Gorg, 2002). Additionally, the experiments revealed that the concentration of COD decreases from 750 mg/L (initial content) to 225 mg/L (final content) at an ideal operational period of 30 min and an ozone formation rate of 1.12 mg/s. The COD degradation efficacy improved from 58.5% to 90.5% after 30 min as the current density was enhanced from 100 to 200 A/m^2, respectively. The effluent's COD content was reduced during electrocoagulation due to the hydroxyl ions formed at the cathodes' surface. Al^{3+} ions were the most prevalent free cation in the solution at low pH levels. The contaminant adsorption into the $Al(OH)_3$ flocs, along with the charge neutralization of colloids were responsible for the degradation of COD in the wastewater (Priya & Jeyanthi, 2019). As the concentration of cationic metal ions rose, so did the amount of contaminants removed. As a result, an enhancement in current density improved the degradation efficacy of COD in the wastewater. Additionally, the results show that the concentration of COD decreases from 750 mg/L (initial content) to 70 mg/L (final content) at the optimum treatment period and current density of 30 min and 200 A/m^2, respectively. Table 8.1 displays the degradation efficacy of electrocoagulation and ozonation processes for the treatment of CRM wastewater

TABLE 8.1

Pollutant Removal Efficiency of Electrocoagulation and Ozonation Processes. (Reproduced with permission from Das, Anweshan, & Purkait, 2021b @ Elsevier)

Parameters	Initial Content	Electrocoagulation (Optimal Conditions)	Removal (%)	Ozonation (Optimal Conditions)	Removal (%)	Desirable Level (WHO)
Phenol (mg/L)	15	0.3 ± 0.1	98	0.95 ± 0.1	93.5	1
Iron (mg/L)	6	0.15 ± 0.2	97.5	0.8 ± 0.2	86.5	1
Oil (mg/L)	25	3 ± 1	88	9.5 ± 0.5	62	10
COD (mg/L)	750	70 ± 2	90.5	225 ± 3	70	250

8.3.2 Evaluation of Operating Cost and Energy Consumption

The cost-effectiveness is examined by calculating the electrical energy per order (E_{EO}) used throughout the lab-scale implementation of the ozonation procedure. E_{EO} is also known as the efficacy index and served as the basis for the figures of merit for technical advancement. The values of E_{EO} in the case of batch reactions were determined in kWh/m³ using the following equation (Bolton, Bircher, Tumas, & Tolman, 2001):

$$E_{EO} = \frac{P \times t \times 1000}{V \times \log\left(\dfrac{C_i}{C_f}\right)}$$

(8.2)

Here, P stands for the required power of the ozonation process (kWh), t stands for operating time (hr), V stands for the sample volume being treated, and C_f and C_i stand for the final and initial pollutant concentration. For ozonation, the values of E_{EO} for the contaminants are: for phenol, E_{EO} = 30.25 kWh/m³; for COD, E_{EO} = 69.32 kWh/m³; for iron E_{EO} = 41.40 kWh/m³; for BOD E_{EO} = 61.4 kWh/m³, and for oil E_{EO} = 86.25 kWh/m³. Likewise, for electrocoagulation, the obtained E_{EO} values are: for phenol E_{EO} = 26.52 kWh/m³; for COD, E_{EO} = 43.70 kWh/m³; for iron E_{EO} = 28.05 kWh/m³; for BOD E_{EO} = 53.25 kWh/m³, for oil E_{EO} = 48.85 kWh/m³.

It becomes abundantly clear that electrocoagulation uses less electrical energy per order than ozonation, proving that this procedure is more effective at removing specific contaminants while using less energy overall. Additionally, the price of energy usage throughout the entire process was taken into account while estimating the operational cost of the ozonation procedure in the current work. The following equation was used to determine the operating cost (Das et al., 2021a):

$$\text{Operating Cost}_{(ozonation)} = q \times Q_{energy}$$

(8.3)

Here, q stands for the price of electricity and Q_{energy} stands for the consumption of electrical energy during pollutant removal. The cost of electricity was calculated based on its price in the state of Assam, India, in 2022. The electrical energy cost was found to be 14.11, 27.50, and 44.75 US$/m³ when taking into account the power utilization of 33, 51, and 72.5 W at 0.85, 1.00, and 1.12 mg/s, respectively. As a result, the operating costs for the formation of ozone at the rates of 0.85, 1.00, and 1.12 mg/s were determined to be 1.36, 2.65, and 4.30 US$/m³, respectively, after a period of 30 min. Sludge disposal costs, fixed costs, chemical costs, electricity costs, and electrode costs make up the operating expenses for electrocoagulation. However, in order to keep things simple, the cost of the electrode and the electricity rate were the key factors in this study's cost assessment. The following equation was used to determine the operational cost of the electrocoagulation procedure (Das, Anweshan, & Purkait, 2021b):

$$\text{Operating Cost}_{(electrocoagulation)} = p \times Q_{electrode} + q \times Q_{energy}$$

(8.4)

Here, Q_{energy} stands for the consumption of electrical energy during pollutant removal, $Q_{electrode}$ stands for the consumption of electrode materials, p stands for the electrode cost, and q stands for the electricity cost. The electrical energy usage is given by the equation below:

$$Q_{energy} = \frac{I \times V \times t}{V_L} \qquad (8.5)$$

Here, I stands for the current (A), V stands for the voltage (V), t stands for the operating time (s), and V_L stands for the sample volume (m^3). Nevertheless, the electrode usage was determined from the following equation (Changmai et al., 2020):

$$Q_{electrode} = \frac{I \times t \times MW}{F \times V_L \times z} \qquad (8.6)$$

Here, z stands for the number of electrons (3) transferred, F stands for the Faraday's constant, and M.W. stands for the molar mass of aluminum. The work conducted demonstrated that an increment in current density raised the price of the electrodes along with the price of energy. Thus, as the current density was enhanced from 100 to 200 A/m^2, the energy cost and the electrode cost varied from 2.82 to 6.5 US\$/m^3 and 0.00106 to 0.00242 US\$/m^3, respectively. For a current density of 100, 150, and 200 A/m^2, and a reaction period of 30 min, the operating cost was computed to be 0.381, 0.553, and 0.742 US\$/m^3, respectively. It was observed that electrocoagulation is more cost-effective than the ozonation process.

8.4 CONCLUSION

For the purpose of treating the phenol, iron, oil and grease, and COD content in actual CRM wastewater from the Tata Steel plant in India, this work concentrates on the utilization of both the electrocoagulation and ozonation procedures. The treatment sample's ability to reduce the pollutants is enhanced by an increment in both the ozone formation rate and current density. Due to the creation of several derivatives with an acidic nature, the ozonation experiment causes the original pH to drop from 7.8 to 3.5. In the course of electrocoagulation, the creation of OH^- ions at the surface of the cathode, however, causes the pH to rise from 7.8 to 8.5. The results of both electrocoagulation and ozonation processes show the optimal operational parameters, viz. 1.12 mg/s (ozone production rate), 200 A/m^2 (current density), and 30 min (operating period) were enough to bring all the contaminant concentrations of CRM wastewater beneath their respective allowable levels, thereby satisfying the environmental requirement for its reuse or/and discharge.

Additionally, it was noted from both tests that, while running at peak efficiency, electrocoagulation outperformed the ozonation procedure in terms of removal effectiveness. Further, it was discovered that the removal effectiveness of the pollutants got slightly enhanced when the ozone production rate was raised above 1.12 mg/s (optimal setting), although a large increment in the process's operating costs was noticed. Additionally, the results of the preliminary analysis showed that for the intended effluent, the operational costs of the ozonation procedure were nearly six times higher than electrocoagulation.

REFERENCES

Abdel-Aziz, M. H., El-Ashtoukhy, E. S. Z., Zoromba, M. S., & Bassyouni, M. (2016). Oil-in-water emulsion breaking by electrocoagulation in a modified electrochemical cell. *International Journal of Electrochemical Science*, *11*(11), 9634–9643. https://doi.org/10.20964/2016.11.53

Adams, C. D., & Gorg, S. (2002). Effect of pH and gas-phase ozone concentration on the decolorization of common textile dyes. *Journal of Environmental Engineering*, *128*(3), 293–298. https://doi.org/10.1061/(asce)0733-9372(2002)128:3(293)

Ashraf, M. I., Ateeb, M., Khan, M. H., Ahmed, N., Mahmood, Q., & Zahidullah. (2016). Integrated treatment of pharmaceutical effluents by chemical coagulation and ozonation. *Separation and Purification Technology*, *158*, 383–386. https://doi.org/10.1016/j.seppur.2015.12.048

Bolton, J. R., Bircher, K. G., Tumas, W., & Tolman, C. A. (2001). Figures-of-merit for the technical development and application of advanced oxidation technologies for both electric- and solar-driven systems. *Pure and Applied Chemistry*, *73*(4), 627–637. https://doi.org/10.1351/pac200173040627

Cerkovnik, J., Eržen, E., Koller, J., & Plesničar, B. (2002). Evidence for HOOO radicals in the formation of alkyl hydrotrioxides (ROOOH) and hydrogen trioxide (HOOOH) in the ozonation of C - H bonds in hydrocarbons. *Journal of the American Chemical Society*, *124*(3), 404–409. https://doi.org/10.1021/ja017320i

Changmai, M., Das, P. P., Mondal, P., Pasawan, M., Sinha, A., Biswas, P., & Purkait, M. K. (2020). Hybrid electrocoagulation–microfiltration technique for treatment of nanofiltration rejected steel industry effluent. *International Journal of Environmental Analytical Chemistry*, *102*(1), 1–22. https://doi.org/10.1080/03067319.2020.1715381

Colla, V., Matino, I., Branca, T. A., Fornai, B., Romaniello, L., & Rosito, F. (2017). Efficient use of water resources in the steel industry. *Water (Switzerland)*, *9*(11), 1–15. https://doi.org/10.3390/w9110874

Das, P. P., Anweshan, Mondal, P., Sinha, A., Biswas, P., Sarkar, S., & Purkait, M. K. (2021a). Integrated ozonation assisted electrocoagulation process for the removal of cyanide from steel industry wastewater. *Chemosphere*, *263*, 128370. https://doi.org/10.1016/j.chemosphere.2020.128370

Das, P. P., Anweshan, & Purkait, M. K. (2021b). Treatment of cold rolling mill (CRM) effluent of steel industry. *Separation and Purification Technology*, *274*(April), 119083. https://doi.org/10.1016/j.seppur.2021.119083

Das, P. P., Mondal, P., Anweshan, Sinha, A., Biswas, P., Sarkar, S., & Purkait, M. K. (2021c). Treatment of steel plant generated biological oxidation treated (BOT) wastewater by hybrid process. *Separation and Purification Technology*, *258*, 118013. https://doi.org/10.1016/j.seppur.2020.118013

Das, P. P., Sharma, M., & Purkait, M. K. (2022). Recent progress on electrocoagulation process for wastewater treatment: A review. *Separation and Purification Technology*, *292*(April), 121058. https://doi.org/10.1016/j.seppur.2022.121058

Doggaz, A., Attour, A., Le Page Mostefa, M., Tlili, M., & Lapicque, F. (2018). Iron removal from waters by electrocoagulation: Investigations of the various physico-chemical phenomena involved. *Separation and Purification Technology*, *203*(April), 217–225. https://doi.org/10.1016/j.seppur.2018.04.045

El Araby, R., Hawash, S., & El Diwani, G. (2009). Treatment of iron and manganese in simulated groundwater via ozone technology. *Desalination*, *249*(3), 1345–1349. https://doi.org/10.1016/j.desal.2009.05.006

Esplugas, S., Giménez, J., Contreras, S., Pascual, E., & Rodríguez, M. (2002). Comparison of different advanced oxidation processes for phenol degradation. *Water Research*, *36*(4), 1034–1042. https://doi.org/10.1016/S0043-1354(01)00301-3

Ghosh, D., Solanki, H., & Purkait, M. K. (2008). Removal of Fe(II) from tap water by electrocoagulation technique. *Journal of Hazardous Materials*, *155*(1–2), 135–143. https://doi.org/10.1016/j.jhazmat.2007.11.042

Khuntia, S., Majumder, S. K., & Ghosh, P. (2015). A pilot plant study of the degradation of brilliant green dye using ozone microbubbles: Mechanism and kinetics of reaction. *Environmental Technology (United Kingdom)*, *36*(3), 336–347. https://doi.org/10.1080/09593330.2014.946971

Kishimoto, N., & Ueno, S. (2012). Catalytic effect of several iron species on ozonation. *Journal of Water and Environment Technology*, *10*(2), 205–215. https://doi.org/10.2965/jwet.2012.205

Krishnan, S., Homroskla, P., Saritpongteeraka, K., Suttinun, O., Nasrullah, M., Tirawanichakul, Y., & Chaiprapat, S. (2022). Specific degradation of phenolic compounds from palm oil mill effluent using ozonation and its impact on methane fermentation. *Chemical Engineering Journal*, *451*(P1), 138487. https://doi.org/10.1016/j.cej.2022.138487

Lee, R., & Coote, M. L. (2016). Mechanistic insights into ozone-initiated oxidative degradation of saturated hydrocarbons and polymers. *Physical Chemistry Chemical Physics*, *18*(35), 24663–24671. https://doi.org/10.1039/c6cp05064f

Mondal, P., & Purkait, M. K. (2017). Green synthesized iron nanoparticle-embedded pH-responsive PVDF-co-HFP membranes: Optimization study for NPs preparation and nitrobenzene reduction. *Separation Science and Technology*, *52*(14), 2338–2355.

Mondal, P., Samanta, N. S., Meghnani, V., & Purkait, M. K. (2019). Selective glucose permeability in presence of various salts through tunable pore size of pH responsive PVDF-co-HFP membrane. *Separation and Purification Technology*, *221*, 249–260.

Mukherjee, R., Mondal, M., Sinha, A., Sarkar, S., & De, S. (2016). Application of nanofiltration membrane for treatment of chloride rich steel plant effluent. *Journal of Environmental Chemical Engineering*, *4*(1), 1–9. https://doi.org/10.1016/j.jece.2015.10.038

Muzyka, K., Sun, J., Fereja, T. H., Lan, Y., Zhang, W., & Xu, G. (2019). Boron-doped diamond: Current progress and challenges in view of electroanalytical applications. *Analytical Methods*, *11*(4), 397–414. https://doi.org/10.1039/c8ay02197j

Oberteuffer, J., Wechsler, I., Marston, P., & Mcnallan, M. (1975). High gradient magnetic filtration of steel mill process and waste waters. *IEEE Transactions on Magnetics*, *M*(5), 1591–1593.

Priya, M., & Jeyanthi, J. (2019). Removal of COD, oil and grease from automobile wash water effluent using electrocoagulation technique. *Microchemical Journal*, *150*(May), 104070. https://doi.org/10.1016/j.microc.2019.104070

Purkait, M. K., Sinha, M. K., Mondal, P., & Singh, R. (2018a). Interface Science and Technology, Elsevier, Pages 115–144.

Purkait, M. K., Sinha, M. K., Mondal, P., & Singh, R. (2018b). Ch 2 - pH-Responsive Membranes, Singh, Interface Science and Technology, Elsevier, Volume 25, Pages 39–66, ISBN 9780128139615.

Purkait, M. K., Sinha, M. K., Mondal, P., & Singh, R. (2018c). Chapter 4 - Photoresponsive Membranes, Editor(s): Mihir Kumar Purkait, Manish Kumar Sinha, Piyal Mondal, Randeep Singh, Interface Science and Technology, Elsevier, Volume 25, Pages 115–144, ISBN 9780128139615.

Purkait, M. K., Mondal, P., & Chang, C.-T. (2019). Treatment of Industrial Effluents: Case Studies, CRC Press, 1st Edition, ISBN 9781138393417.

Purkait, M. K., Singh, R., Mondal, P., & Haldar, D. (2020). Thermal Induced Membrane Separation Processes, Elsevier, ISBN 9780128188019.

Ramseier, M. K., & von Gunten, U. (2009). Mechanisms of phenol ozonation-kinetics of formation of primary and secondary reaction products. *Ozone: Science and Engineering*, *31*(3), 201–215. https://doi.org/10.1080/01919510902740477

Rekhate, C. V., & Srivastava, J. K. (2020). Recent advances in ozone-based advanced oxidation processes for treatment of wastewater - A review. *Chemical Engineering Journal Advances*, *3*(June), 100031. https://doi.org/10.1016/j.ceja.2020.100031

Samanta, N. S., Banerjee, S., Mondal, P., Anweshan, Bora, U., & Purkait, M. K. (2021). Preparation and characterization of zeolite from waste Linz-Donawitz (LD) process slag of steel industry for removal of Fe^{3+} from drinking water. *Advanced Powder Technology*, *32*(9), 3372–3387.

Sharma, M., Das, P. P., Sood, T., Chakraborty, A., & Purkait, M. K. (2021). Ameliorated polyvinylidene fluoride based proton exchange membrane impregnated with graphene oxide, and cellulose acetate obtained from sugarcane bagasse for application in microbial fuel cell. *Journal of Environmental Chemical Engineering*, *9*(6), 106681. https://doi.org/10.1016/j.jece.2021.106681

Sharma, M., Das, P. P., Sood, T., Chakraborty, A., & Purkait, M. K. (2022). Reduced graphene oxide incorporated polyvinylidene fluoride/cellulose acetate proton exchange membrane for energy extraction using microbial fuel cells. *Journal of Electroanalytical Chemistry*, *907*(November 2021), 115890. https://doi.org/10.1016/j.jelechem.2021.115890

Symons, C. R. (1971). Treatment of cold-mill wastewater by ultra-high-rate filtration. *Water Pollution Control Federation*, *43*(11), 2280–2286. https://doi.org/doi.org/10.1109/WPCF.1975.101453

Uğurlu, M., Gürses, A., Doğar, Ç, & Yalçin, M. (2008). The removal of lignin and phenol from paper mill effluents by electrocoagulation. *Journal of Environmental Management*, *87*(3), 420–428. https://doi.org/10.1016/j.jenvman.2007.01.007

von Sonntag, C., & von Gunten, U. (2015). Chemistry of Ozone in Water and Wastewater Treatment: From Basic Principles to Applications, Chemistry of Ozone in Water and Wastewater Treatment: From Basic Principles to Applications. https://doi.org/10.2166/9781780400839

Yang, L. P., Hu, W. Y., Huang, H. M., & Yan, B. (2010). Degradation of high concentration phenol by ozonation in combination with ultrasonic irradiation. *Desalination and Water Treatment*, *21*(1–3), 87–95. https://doi.org/10.5004/dwt.2010.1233

Zhang, Y., Gan, F., Li, M., Li, J., Li, S., & Wu, S. (2010). New integrated processes for treating cold-rolling mill emulsion wastewater. *Journal of Iron and Steel Research International*, *17*(6), 32–35. https://doi.org/10.1016/S1006-706X(10)60110-0

9 Sustainable Management, Techno-Economic Analysis, and Future Perspectives

9.1 INTRODUCTION

The global demand for electricity is increasing rapidly. The electricity supply sector may be unable to keep up with rising demand due to the high costs of increasing generation capacity and expanding transmission/distribution systems. As a result, energy shortages and peak demand excesses may occur during critical peak conditions, forcing customers to face involuntary load shedding. Electric energy consumption by industrial loads accounts for a significant portion of the total energy produced. The industrial sector consumes approximately 41% of total electrical energy in most countries (Abdelaziz, Saidur, & Mekhilef, 2011).

9.2 SUSTAINABLE MANAGEMENT OF STEEL INDUSTRY WASTEWATER

Metal recovery from effluent from the steel industry has been successfully accomplished using a number of technologies, including the electrochemical process, adsorption process, membrane process, and hybrid process. For instance, Rodrigues, Marder, Moura Bernardes, and Zoppas Ferreira (2014) extracted cadmium cyanide complexes and CN^- from electroplating wastewater using electro-dialysis. The greatest extraction for cadmium cyanide complexes and CN^-, respectively, was 20.1% and 44.1% at the end of 2 hours, according to the results. In addition, it was shown that the recovery rate fell when the molar ratio between the concentration of contaminant and metal ions was decreased below its ideal level.

San Roman, Ortiz, Andara, Ibanez, and Ortiz (2012) separated and recovered Fe, Zn, and HCl from pickling wastewater using a hybrid membrane approach. The technique incorporates diffusion dialysis and emulsion pertraction for the separation and recovery of metals. Ramirez et al. (2020) created a revolutionary method for denitrifying wastewater made of stainless steel called an anaerobic swirling fluidized bed membrane bioreactor. According to the findings, more than 94% of the nitrate had been eliminated, and 40% of the wastewater's metals (Fe, Cr, Ni, Co, etc.) had been recovered. Jiang, Li, Xu, and Ruan (2021) tested a hybrid

DOI: 10.1201/9781003366263-9

treatment system that included electro-dialysis, RO, and bipolar membrane electro-dialysis to handle cold-rolling wastewater. The hybrid system worked well and had the ability to recover salts and acid/base from wastewater from cold rolling.

Additionally, a combination of CW, UF, and RO was tested to treat wastewater from the steel sector in order to lessen freshwater use and reclaim mixed treated wastewater. The hybrid system demonstrated 75% metal recovery and 98% salt removal from wastewater (Huang, Ling, Xu, Feng, & Li, 2011). In the steel industry, a lot of steel slag is produced as trash or byproducts from various furnaces (Lim, Chew, Choong, Tezara, & Yazdi, 2016). Oxides of calcium, iron, silicon, aluminum, and other elements are found in steel slag. According to the US Geological Survey (2020), the price of steel industry slag in the United States was predicted to be $27.65 per ton in 2019.

BF slag dominates the market, accounting for 88% of the entire value and around 50% of the total quantity. The Indian steel industry produces roughly 24 million tons of slag yearly, according to the Indian Ministry of Mines (2020). Furthermore, by 2030, the amount of BF slag production could reach 50 million tons. The slag's innate physicochemical characteristics make it a viable adsorbent for wastewater treatment. Steel slag, such as BF slag, BOF slag, and EAF slag, has recently been widely employed as an adsorbent to remove pollutants from wastewater. Slag can also be used as a building material, an active filter media, and a raw ingredient in the production of cement.

9.3 TECHNO-ECONOMIC ANALYSIS OF TREATMENT PROCESSES

A few scholars have in the past created economic models to represent the wastewater treatment systems used in the steel sector. To determine the most effective and environmentally friendly wastewater treatment method for the steel industry, Mahjouri et al. (2017) developed an integrated methodology (Analytic Hierarchy Process and Technique for order preference by similarity to ideal solution methods with fuzzy logic as a hybrid Fuzzy multi-criteria decision-making). Due to its higher efficacy, economic suitability, and compliance with environmental regulations, the results demonstrated that the combination of advanced treatment (electrolytic splitting with RO and evaporation) and conventional treatment (neutralization, flocculation, sedimentation, flotation, and filtration) was highly effective.

The entire running cost for wastewater treatment plants includes both direct and indirect costs, such as labor, maintenance, and depreciation of fixed assets. Direct costs include power, materials (chemicals), and sludge management (Kara, 2013). In general, an improvement in effluent quality leads to an increase in WWTP's operating costs. Primary, secondary, and tertiary treatments had respective mean operational costs of 0.12, 0.26, and 0.32 €/m³ (Molinos-senante, Hernández-sancho, & Sala-garrido, 2010). Lim et al. (2016) used LCA to predict the environmental and economic likelihood of TWTNS in their study. The current steel industry's WWTPs were networked, and data including flow rates, pollutants concentration (COD, SS, and F⁻), removal efficacy, and plant capacity were taken into account for network synthesis.

Additionally, a comparison of TWTNS and CWTS revealed that TWTNS had a greater environmental impact (29.6–68.3%) than CWTS, which could be attributed to the networked system's higher power consumption. However, TWTNS was also found to be more cost-effective due to the networked WWTPs' lower labor costs. According to Tihansky's (1972) analysis of the expenses associated with waste management in the steel industry, the expenditures for the first 95% of treatment and the final 5% of treatment are often equal. Colla et al. (2017) looked into the economic viability of using UF and RO techniques to treat steel industry effluent and sustainably reusing treated water. Energy, maintenance, chemicals, membranes, and labor were determined to be the key contributions to the total operating expenses (0.22 €/m^3). According to Das et al. (2021a), the operational cost (5.801 US$/m^3) for the combined electro-coagulation and ozonation processes was the sum of the energy requirements for both processes and the electrode cost for electro-coagulation, which was comparable with hybrid systems mentioned elsewhere. Since steel production uses a lot of water, it's important to consider ways to cut back on freshwater usage. In the steel industry, Zhang, Zhao, Cao, and Wen (2018) proposed models based on the integration of water networks between and within plants. The models were successfully applied to reduce the freshwater consumption and the overall network by 22% and 20%, respectively.

The cost of treating NF reject water with the RO process and solvent-based precipitation was $7.35/m^3. A field-scale implementation of an integrated CN$^-$ removal method like UV-H$_2$O$_2$ was found to cost $4.63 per m^3 of wastewater that was treated per hour. Despite their effectiveness, additional research is needed to fairly estimate the cost of implementation while taking into account all cost-affecting aspects (Deepti et.al., 2021).

9.4   COST ESTIMATION STUDY OF WASTEWATER TREATMENT PROCESSES

The ammoniacal liquor from coke ovens contains phenol (an industrial solvent), ammonia (a raw material for fertilizers), and cyanide (an ingredient in the production of paper, textiles, and plastics), all of which have been recovered successfully. Table 9.1 illustrates how value-added products were profitably recovered from wastes generated by various units of a steel plant. However, evaluating process economics is critical when assessing recovery options. Furthermore, capital, installation, and maintenance costs must be considered, as well as the cost of raw materials and production, the volume of effluent generated, and the final disposal standards established by environmental protection agencies. Tihansky (1972) examined the total cost of pollution control incurred by all American steel plants.

It was discovered that the cost of installation is determined by the volume of effluent and the efficiency required. It rises when removal efficiencies exceed a certain threshold limit. In contrast to the widely held belief that reverses osmosis membrane filtration processes are more expensive (Mondal & Purkait, 2017; Mondal, Samanta, Meghnani, & Purkait, 2019; Purkait, Mondal, & Chang, 2019; Purkait, Singh, Mondal, & Haldar, 2020). Ari, Ozgun, Ersahin, and Koyuncu

TABLE 9.1

Profitable By-Product Recovery. (Data from Das et al., 2018, Copyright © John Wiley and Sons)

Waste	Component	Product	Recovery method
Ammoniacal Liquor from Coke Ovens	Ammonia	i. Struvite (a slow-releasing fertilizer) ii. Nitrate	i. Precipitation with magnesium and manganese ores ii. Biological treatment
Blast Furnace Flue Gas	Carbon	Iron-rich grains	Froth flotation, followed by low-intensity magnetic separation
Converter Slag	Magnetite	Adsorbent for nickel adsorption	Slurry filtration
Mixed Pond Ash from Captive Power Plants	–	Bricks	Hydration
Metallurgical Dust from Iron and Steel Plants	Iron, zinc, carbon	Pellets, mud, and slurry	Borne dressings, gravity separation, leaching, and molding technology

(2011) demonstrated that the operating costs of reverse osmosis membrane filtration processes are nearly identical to those of nano-filtration.

In order to study the performance of water resource recovery facilities, the two most important parameters for assessing techno-economic feasibility are WPI (Water Price Index) and Energy Consumption Curves. Kumar, Groth, and Vlacic (2015) used the following formula to calculate the Water Price Index:

$$Water\ Price\ Index\ (\$/KL) = (Annual\ Plant\ Cost\ /\ Plant\ Performance) \times Margin \tag{9.1}$$

The annual costs can be calculated by the addition of depreciation and expenses incurred during the operation of the equipment:

$$Plant\ Cost\ (in\ \$) = Depreciation + Operational\ expenditure \tag{9.2}$$

For waste treatment plants with "n" number of different components where the rate of depreciation is different, the total depreciation cost of each component "i" was computed with the help of the following equation:

$$Depreciation\ (\$/year) = \sum_{i=1}^{n} \frac{Capital\ cost\ of\ instruments}{Projected\ life\ of\ instruments} \tag{9.3}$$

Energy consumption curves, in this case, "cost of conserved energy" (CCE), can be constructed to rate the energy efficiency of a technology, taking into account the cost of conserved energy to balance the total cost of new technology and the energy savings obtained from it (Worrell, Price, & Martin, 2001). The CCE can be calculated using the following formula:

$$CCE = \frac{Annual\ Investment + Annual\ change\ in\ O\&M\ costs}{Annual\ Energy\ Savings} \quad (9.4)$$

The total annualized investment at a discount rate "d" for the duration of "n" periods can be evaluated using the following equation:

$$Annualized\ investment = Capital\ Cost \times \frac{d}{\left[1-(1+d)^{-n}\right]} \quad (9.5)$$

Another method for lowering costs and increasing water reclamation is redesigning the treatment while introducing new production technologies. Using dry quenching instead of wet quenching reduces energy costs by nearly 40% while also reducing the amount of hazardous ammoniacal liquor produced. New steel plants are increasingly avoiding coke ovens in favor of direct reduction processes like HISMELT, which involves the direct smelting of iron ore, lime, and coal in a smelt reduction vessel (SRV). This reduces process costs by allowing the use of non-coking coal and Titano-magnetite iron ore, both of which are available in India, without compromising steel quality.

The economic aspect of the ozonation process is investigated during the lab scale (batch operation) application by evaluating electrical energy per order (E_{EO}). The figures of merit for technical development were calculated using the electrical energy per order (E_{EO}) (also known as the efficiency index).

E_{EO} is the amount of electrical energy required in kWh to reduce pollutant content in 1 m^3 of contaminated water by one order of magnitude (Bolton, Bircher, Tumas, & Tolman, 2001).

The E_{EO} (kWh/m^3) values for batch operations for pseudo-first-order reactions were calculated using the following equation:

$$E_{EO} = \frac{P \times t \times 1000}{V \times \log_{10}\left(\dfrac{C_i}{C_f}\right)} \quad (9.6)$$

where P = power required for the process (kWh), t = reaction time (hr), V = volume of sample treated, and C_i and C_f are the initial and final concentrations of the pollutants. In ozonation, the E_{EO} values for the pollutants are as follows: for COD, E_{EO} = 69.32 kWh m^{-3}; for BOD, E_{EO} = 61.4 kWh m^{-3}; for phenol, E_{EO} = 30.25 kWh m^{-3}; for iron, E_{EO} = 41.40 kWh m^{-3}; for oil, E_{EO} = 86.25 kWh m^{-3}. Similarly, in the case of electrocoagulation, the values of E_{EO} obtained are as follows: for COD, E_{EO} = 43.70 kWh m^{-3}; for BOD, E_{EO} = 53.25 kWh m^{-3}; for phenol, E_{EO} = 26.52 kWh m^{-3}; for iron, E_{EO} = 28.05 kWh m^{-3}; for oil, E_{EO} = 48.85 kWh m^{-3}. The electrical energy per order for electrocoagulation is clearly lower than that of ozonation, indicating that the electrocoagulation process is more energy-efficient in terms of removing specified pollutants.

Furthermore, in the current study, the cost of ozonation was primarily determined by the cost of energy required throughout the entire operation (US$m^{-3} of solution). The operational cost was calculated using the following equation:

$$Operating\ cost_{(ozonation)} = q \times Q_{energy} \qquad (9.7)$$

Here, q = Electricity cost (0.0948 US$kWh^{-1}) and Q_{energy} = Electrical energy consumption for pollutant removal. The electricity cost was considered in accordance with its price for Assam (India) in the year 2020. Considering the power utilization to be 33, 51, and 72.5 W at 0.85, 1.00, and 1.12 mg s^{-1}, the price of electrical energy was observed to be 14.11, 27.50, and 44.75 US$m^{-3}, respectively. As such, the cost of operation for the ozone generation rate of 0.85, 1.00, and 1.12 mg s^{-1} was calculated as 1.36, 2.65, and 4.30 US$m^{-3}, respectively, at the end of 30 min of treatment time.

The cost of operation for electrocoagulation consists of sludge disposal cost, fixed cost, cost of chemicals, electricity, and electrode cost. Nevertheless, for simplicity, the cost estimation in this study mainly involved the rate of electricity and the electrode cost. The operating cost was determined from the equation given below (Changmai, Pasawan, & Purkait, 2018):

$$Operating\ cost_{(electrocoagulation)} = p \times Q_{electrode} + q \times Q_{energy} \qquad (9.8)$$

Here, Q_{energy} = Electrical energy consumption for pollutant removal, $Q_{electrode}$ = Electrode materials consumption, p = Cost of electrode (2.0475 US$kg^{-1} of aluminum) and q = Cost of electricity (0.0948 US$kWh^{-1}). The equation for electrical energy consumption is given as follows:

$$Q_{energy} = \frac{I \times V \times t}{V_L} \qquad (9.9)$$

Here, I = Current (A), V = Voltage (V), t = treatment time (s) and V_L = Wastewater volume (m^3). However, the consumption of electrodes was evaluated from the equation below (Faraday's law):

$$Q_{electrode} = \frac{I \times t \times M.W}{F \times V_L \times z} \qquad (9.10)$$

Here, M.W = Molar mass of aluminum (26.98 g mol^{-1}), F = Faraday's constant (96,487 C mol^{-1}) and z = Number of transferred electrons (z = 3). The work done showed that, with an enhanced current density, the cost of energy and the electrodes increased. The reason can be attributed to higher anodic oxidation and energy consumption during the process. Thus, with a rise in current density from 100 to 200 A m^{-2}, the cost of electrodes as well as the cost of energy varied from 0.00106 to 0.00242 US$m^{-3} and 2.82 to 6.5 US$m^{-3}, respectively. The operating cost as such was calculated to be 0.381, 0.553, and 0.742 US$m^{-3} for a current density of 100, 150, and 200 A m^{-2}, respectively, after a reaction time of 30 min. Figure 9.1 shows the cost of operation for both ozonation and electrocoagulation

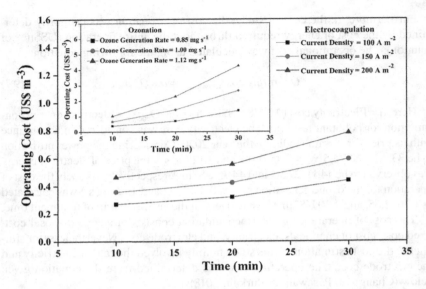

FIGURE 9.1 Performance assessment of both the processes on operating cost: Inset: ozonation process, Outset: electrocoagulation process. (Reproduced with permission from Das, Anweshan, & Purkait, 2021b, Copyright © Elsevier.)

processes. From Figure 9.1, it is seen that electrocoagulation is much more economical compared to the ozonation process. The mentioned operational cost was found to be economical both for electrocoagulation as well as ozonation when compared with other reported wastewater treatment methods, as shown in Table 9.2 (Canizares, Paz, Sáez, & Rodrigo, 2009; Coimbra et al., 2021; El-Dein,

TABLE 9.2
Comparison of Cost Estimation of Ozonation and Electrocoagulation Processes for Different Types of Wastewater (Reproduced with permission from Das, Anweshan, & Purkait, 2021b, Copyright © Elsevier)

Methods used	Types of Wastewater	Operating Cost
Electrocoagulation	Metalworking wastewater	4.74 US$ m^{-3}
Electrocoagulation	Synthetic wastewater	1.62 US$ m^{-3}
Electrocoagulation	Kraft pulp bleaching filtrates	Acid and Alkaline filtrate = 12.59 and 14.21 US$ m^{-3}
Electrocoagulation	Synthetic wastewater	6.05 US$ m^{-3}
Ozonation	Microalgae lipid strain slurry	£5.76 ≅ 7.44 US$ (kg of dry algal mass)$^{-1}$
Ozonation	Textile dye wastewater	54 € m^{-3} ≅ 63.5 US$ m^{-3}
Ozonation	Olive oil mill wastewater	Olive oil mill wastewater = 81 €/m^3 ≅ 95 US$/m^3
Ozonation	Municipal wastewater	Pre- and Post-ozonation reactor = 4.50 and 3.90 US$ m^{-3}

Libra, & Wiesmann, 2006; Irani, Khoshfetrat, & Forouzesh, 2021; Kamaroddin, Rahaman, Gilmour, & Zimmerman, 2020; Kobya, Omwene, & Ukundimana, 2020; Madhavan & Antony, 2021)

9.5 CONCLUSION AND FUTURE RECOMMENDATIONS

The steel industry generates a large amount of wastewater, which necessitates the selection of appropriate treatment technologies based on various parameters such as source, composition, contaminant concentration, and discharge standards. Conventional biological treatment techniques appear capable of treating large volumes; however, improved biodegradability of recalcitrant pollutants is required for efficient treatment. Furthermore, because the concentration of contaminants in steel effluent is high, single-step processes necessitate a longer retention time, which may result in toxic effects of CN^- and phenol on nitrifying bacteria. Sequencing of reactors/phases (Anoxic/Oxic/Anaerobic) would aid in achieving a diverse range of microorganisms in this regard. It would improve the biodegradability of PAHs and CN^- while also effectively removing nitrogen compounds. Furthermore, a thorough assessment of the microbial community and optimization of operational parameters can lead to the establishment of a large-scale relationship between system stability and biological functions.

AOPs are clearly effective in treating recalcitrant contaminants; for example, combining AOPs such as ozone with EC, UV, and H_2O_2 can remove contaminants below permissible limits. However, because these techniques are energy-intensive, require a large amount of chemicals, and have high application costs, their industrial application is difficult. The majority of recent studies have been conducted on a lab scale, with only a few field-scale investigations. As a result, more field research is needed, focusing on technology integration, cost implications, and secondary pollution.

This focus book emphasizes that the attention of the steel industry's effluent treatment should not be limited to meeting discharge standards but to achieving reusable quality water. Membrane technologies provide an adequate solution in this context, and their selection should be based on the potential use of water, energy consumption, and cost. Membrane fouling, cost, and replacement issues can be addressed by developing new membrane materials with low fouling potential and integrating membranes with other processes. As a result, with the increasing demand for and cost of freshwater, more research focusing on membranes as a standalone and integrated solution is critical.

Furthermore, long-term strategies for stabilizing membrane concentrates prior to disposal can reduce the risk of secondary pollution. Utilization of steel industry byproducts such as slag and other waste materials such as rice husk as adsorbents can be critical (Purkait, Sinha, Mondal, & Singh, 2018a, b, c). Future research should concentrate on the development, modification, regeneration, and long-term management of spent adsorbents. Proper water and wastewater networking, as well as determining the best low-cost treatment technology for specific contaminated streams, can also provide an effective solution.

Researchers in this field can gain valuable insights into the factors governing the optimal performance of emerging techniques for the treatment of recalcitrant pollutants, particularly CN⁻ and phenols (Changmai et al., 2022; Samanta et al., 2021). This article also assesses the critical experimental parameters that will aid in overcoming the limitations associated with existing and emerging technologies. The importance of key aspects determining the effectiveness of integrated systems is highlighted, which will assist industrial practitioners in establishing a sustainable treatment chain to treat a concoction of contaminants originating at various stages of steel manufacturing.

It is important to note that the suggested solutions are intended to encourage energy conservation and effluent reduction in the steel sector and are essential for industrial practitioners to establish successful development strategies from the perspective of the circular economy. To increase resource recovery and develop water footprint reduction methods explained in earlier sections, stakeholders and practitioners must concentrate on LCA.

This study will provide new motivation for researchers and industry personnel to adopt appropriate integrated technology and establish research and development facilities with a focus on the utilization of waste as a resource. The management of effluents in the steel industry is a complex and challenging task.

REFERENCES

Abdelaziz, E. A., Saidur, R., & Mekhilef, S. (2011). A review on energy saving strategies in industrial sector. *Renewable and Sustainable Energy Reviews, 15*(1), 150–168.

Ari, P. H., Ozgun, H., Ersahin, M. E., & Koyuncu, I. (2011). Cost analysis of large scale membrane treatment systems for potable water treatment. *Desalination and Water Treatment, 26*, 172–177.

Bolton, J. R., Bircher, K. G., Tumas, W., & Tolman, C. A. (2001). Figures-of-merit for the technical development and application of advanced oxidation technologies for both electric- and solar-driven systems (IUPAC technical report). *Pure and Applied Chemistry, 73*, 627–637.

Canizares, P., Paz, R., Sáez, C., & Rodrigo, M. A. (2009). Costs of the electrochemical oxidation of wastewaters: A comparison with ozonation and Fenton oxidation processes. *Journal of Environmental Management, 90*, 410–420.

Changmai, M., Pasawan, M., & Purkait, M. K. (2018). A hybrid method for the removal of fluoride from drinking water: Parametric study and cost estimation. *Separation and Purification Technology, 206*, 140–148.

Changmai, M., Das, P. P., Mondal, P., Pasawan, M., Sinha, A., Biswas, P., Sarkar, S., & Purkait, M. K. (2022). Hybrid electrocoagulation–microfiltration technique for treatment of nanofiltration rejected steel industry effluent. *International Journal of Environmental Analytical Chemistry, 102*(1), 62–83.

Coimbra, E. C. L., Mounteer, A. H., do Carmo, A. L. V., Michielsen, M. J. F., Tótola, L., Alcântara, G., Júlia, P. F., Gonçalves, Júlia, G. A. N., & da Silva, P. R. (2021). Electrocoagulation of kraft pulp bleaching filtrates to improve bio-treatability, process. *Safety and Environmental Protection, 147*, 346–355.

Colla, V., Matino, I., Branca, T., Fornai, B., Romaniello, L., & Rosito, F. (2017). Efficient use of water resources in the steel industry. *Water, 9*, 874.

Das, P. P., Anweshan, Mondal, P., Sinha, A., Biswas, P., Sarkar, S., & Purkait, M. K. (2021a). Integrated ozonation assisted electrocoagulation process for the removal of cyanide from steel industry wastewater. *Chemosphere*, *263*, 128370.

Das, P. P., Anweshan, & Purkait, M. K. (2021b). Treatment of cold rolling mill (CRM) effluent of steel industry. *Purification Technology*, *274*, 119083.

Das, P., Mondal, G. C., Singh, S., Singh, A. K., Prasad, B., & Singh, K. K. (2018). Effluent treatment technologies in the iron and steel industry - a state of the art review. *Water Environment Research*, *90*, 395–408.

Deepti., Bora, U., & Purkait, M. K. (2021). Promising integrated technique for the treatment of highly saline nanofiltration rejected stream of steel industry. *Journal of Environmental Management*, *300*, 113781.

Huang, X.-F., Ling, J., Xu, J.-C., Feng, Y., & Li, G.-M. (2011). Advanced treatment of wastewater from an iron and steel enterprise by a constructed wetland/ultrafiltration/reverse osmosis process. *Desalination*, *269*, 41–49.

Indian Ministry of Mines. (2020). Indian Minerals Yearbook. Indian Bureau of Mines.

Irani, R., Khoshfetrat, A. B., & Forouzesh, M. (2021). Real municipal wastewater treatment using simultaneous pre and post-ozonation combined biological attached growth reactor: Energy consumption assessment. *The Journal of Environmental Chemical Engineering*, *9*(1), 104595.

Jiang, G., Li, H., Xu, M., & Ruan, H. (2021). Sustainable reverse osmosis, electro-dialysis and bipolar membrane electro-dialysis application for cold-rolling wastewater treatment in the steel industry. *The Journal of Water Process Engineering*, *40*, 101968.

Kamaroddin, M. F., Rahaman, A., Gilmour, D. J., & Zimmerman, W. B. (2020). Optimization and cost estimation of microalgal lipid extraction using ozone-rich microbubbles for biodiesel production. *Biocatalysis and Agricultural Biotechnology*, *23*, 101462.

Kara, S. (2013). Treatment of transport container washing wastewater by electrocoagulation. *Environmental Progress & Sustainable Energy*, *32*, 249–256.

Kobya, M., Omwene, P. I., & Ukundimana, Z. (2020). Treatment and operating cost analysis of metalworking wastewaters by a continuous electrocoagulation reactor. *The Journal of Environmental Chemical Engineering*, *8*(2), 103526.

Kumar, S., Groth, A., & Vlacic, L. (2015). Cost evaluation of water and wastewater treatment plants using water price index. *Water Resources Management*, *29*, 3343–3356.

Lim, J. W., Chew, L. H., Choong, T. S. Y., Tezara, C., & Yazdi, M. H. (2016). Overview of steel slag application and utilization. *MATEC Web of Conferences*, *74*, 00026.

Madhavan, M. A., & Antony, S. P. (2021). Effect of polarity shift on the performance of electrocoagulation process for the treatment of produced water. *Chemosphere*, *263*, 128052.

Mahjouri, M., Ishak, M. B., Torabian, A., Abd Manaf, L., Halimoon, N., & Ghoddusi, J. (2017). Optimal selection of iron and steel wastewater treatment technology using integrated multi-criteria decision-making techniques and fuzzy logic. *Process Safety and Environmental Protection*, *107*, 54–68.

Molinos-Senante, M., Hernández-Sancho, F., & Sala-Garrido, R. (2010). Science of the total environment economic feasibility study for wastewater treatment: A cost – Benefit analysis. *Science of the Total Environment*, *408*, 4396–4402.

Mondal, P., & Purkait, M. K. (2017). Green synthesized iron nanoparticle-embedded pH-responsive PVDF-co-HFP membranes: Optimization study for NPs preparation and nitrobenzene reduction. *Separation Science and Technology*, *52*(14), 2338–2355.

Mondal, P., Samanta, N. S., Meghnani, V., & Purkait, M. K. (2019). Selective glucose permeability in presence of various salts through tunable pore size of pH responsive PVDF-co-HFP membrane. *Separation and Purification Technology*, *221*, 249–260.

Purkait, M. K., Sinha, M. K., Mondal, P., & Singh, R. (2018a). Interface Science and Technology, Elsevier, Pages 115–144.

Purkait, M. K., Sinha, M. K., Mondal, P., & Singh, R. (2018b). Ch 2 - pH-Responsive Membranes, Singh, Interface Science and Technology, Elsevier, Volume 25, Pages 39–66, ISBN 9780128139615.

Purkait, M. K., Sinha, M. K., Mondal, P., & Singh, R. (2018c). Chapter 4 - Photoresponsive Membranes, Editor(s): M. K. Purkait, M. K. Sinha, P. Mondal & R. Singh, Interface Science and Technology, Elsevier, Volume 25, Pages 115–144, ISBN 9780128139615.

Purkait, M. K., Mondal, P., & Chang, C.-T. (2019). Treatment of Industrial Effluents: Case Studies, CRC Press, 1st Edition, ISBN 9781138393417.

Purkait, M. K., Singh, R., Mondal, P., & Haldar, D. (2020). Thermal Induced Membrane Separation Processes, Elsevier, ISBN 9780128188019.

Ramírez, J. E., Esquivel-Gonzalez, S., Rangel-Mendez, J. R., Arriaga, S. L., Gallegos-García, M., Buitron, G., & Cervantes, F. J. (2020). Bio-recovery of metals from a stainless steel industrial effluent through denitrification performed in a novel anaerobic swirling fluidized membrane bioreactor (ASFMBR). *Industrial & Engineering Chemistry Research*, 59, 2725–2735.

Rodrigues, M. A. S., Marder, L., Moura Bernardes, A., & Zoppas Ferreira, J. (2014). Electrodialysis Treatment of Metal-Cyanide Complexes, *Electrodialysis and Water Reuse*, Springer Berlin Heidelberg, Berlin, Heidelberg, Pages 119–131.

Samanta, N. S., Banerjee, S., Mondal, P., Anweshan, Bora, U., & Purkait, M. K. (2021). Preparation and characterization of zeolite from waste Linz-Donawitz (LD) process slag of steel industry for removal of Fe^{3+} from drinking water. *Advanced Powder Technology*, 32(9), 3372–3387.

San Roman, M. F., Ortiz, G., Andara, I., Ibanez, R., & Ortiz, I. (2012). Hybrid membrane process for the recovery of major components (zinc, iron and HCl) from spent pickling effluents. *Journal of Membrane Science*, 415–416, 616–623.

Tihansky, D. P. (1972). A cost analysis of waste management in the steel industry. *Journal of the Air Pollution Control Association*, 22, 335–341.

US. Geological Survey. (2020). Mineral Commodities Summary. 86–87.

Worrell, E., Price, L., & Martin, N. (2001). Energy efficiency and carbon dioxide emissions reduction opportunities in the US iron and steel sector. *Ener*, 26, 513–536.

Zhang, K., Zhao, Y., Cao, H., & Wen, H. (2018). Multi-scale water network optimization considering simultaneous intra- and inter-plant integration in steel industry. *Journal of Cleaner Production*, 176, 663–675.

Index

Printed in the United States
by Baker & Taylor Publisher Services

Printed in the United States
by Baker & Taylor Publisher Services